SCIENCE AND TECHNOLOGY FOR ENVIRONMENTAL CLEANUP AT HANFORD

Committee on the Review of the Hanford Site's
Environmental Remediation Science and Technology Plan

Board on Radioactive Waste Management

Division on Earth and Life Studies

National Research Council

NATIONAL ACADEMY PRESS
Washington, D.C.

NOTICE: The project that is the subject of this report was approved by the Governing Board of the National Research Council, whose members are drawn from the councils of the National Academy of Sciences, the National Academy of Engineering, and the Institute of Medicine. The members of the committee responsible for the report were chosen for their special competencies and with regard for appropriate balance.

Support for this study was provided by the U.S. Department of Energy under cooperative agreement number DE-FC01-99EW59049. All opinions, findings, conclusions, and recommendations expressed herein are those of the authors and do not necessarily reflect the views of the Department of Energy.

International Standard Book Number 0-309-07596-3

Additional copies of this report are available from:

National Academy Press
2101 Constitution Avenue, N.W.
Box 285
Washington, DC 20055
800-624-6242
202-334-3313 (in the Washington Metropolitan Area)
http://www.nap.edu

Copyright 2001 by the National Academy of Sciences. All rights reserved.

Printed in the United States of America.

THE NATIONAL ACADEMIES

National Academy of Sciences
National Academy of Engineering
Institute of Medicine
National Research Council

The **National Academy of Sciences** is a private, nonprofit, self-perpetuating society of distinguished scholars engaged in scientific and engineering research, dedicated to the furtherance of science and technology and to their use for the general welfare. Upon the authority of the charter granted to it by the Congress in 1863, the Academy has a mandate that requires it to advise the federal government on scientific and technical matters. Dr. Bruce Alberts is president of the National Academy of Sciences.

The **National Academy of Engineering** was established in 1964, under the charter of the National Academy of Sciences, as a parallel organization of outstanding engineers. It is autonomous in its administration and in the selection of its members, sharing with the National Academy of Sciences the responsibility for advising the federal government. The National Academy of Engineering also sponsors engineering programs aimed at meeting national needs, encourages education and research, and recognizes the superior achievements of engineers. Dr. William A. Wulf is president of the National Academy of Engineering.

The **Institute of Medicine** was established in 1970 by the National Academy of Sciences to secure the services of eminent members of appropriate professions in the examination of policy matters pertaining to the health of the public. The Institute acts under the responsibility given to the National Academy of Sciences by its congressional charter to be an adviser to the federal government and, upon its own initiative, to identify issues of medical care, research, and education. Dr. Kenneth Shine is president of the Institute of Medicine.

The **National Research Council** was organized by the National Academy of Sciences in 1916 to associate the broad community of science and technology with the Academy's purposes of furthering knowledge and of advising the federal government. Functioning in accordance with general policies determined by the Academy, the Council has become the principal operating agency of both the National Academy of Sciences and the National Academy of Engineering in providing services to the government, the public, and the scientific and engineering communities. The Council is administered jointly by both Academies and the Institute of Medicine. Dr. Bruce Alberts and Dr. Wm. A. Wulf are chairman and vice-chairman, respectively, of the National Research Council.

COMMITTEE ON THE REVIEW OF THE HANFORD SITE'S ENVIRONMENTAL REMEDIATION SCIENCE AND TECHNOLOGY PLAN

CHRIS G. WHIPPLE, *Chair*, ENVIRON International Corporation, Emeryville, California
D. WAYNE BERMAN, Aeolus, Inc., Albany, California
SUE B. CLARK, Washington State University, Pullman
JOHN C. FOUNTAIN, State University of New York, Buffalo
LYNN W. GELHAR, Massachusetts Institute of Technology, Cambridge
LISA C. GREEN, Lucent Technologies, Norcross, Georgia
ROBERT O. HALL, University of Wyoming, Laramie
EDWIN E. HERRICKS, University of Illinois, Urbana
BRUCE D. HONEYMAN, Colorado School of Mines, Golden
SALOMON LEVY, Levy & Associates, San Jose, California
JAMES K. MITCHELL, Virginia Polytechnic Institute and State University (retired), Blacksburg
LEON T. SILVER, California Institute of Technology (retired), Pasadena
LESLIE SMITH, University of British Columbia, Vancouver
DAVID A. STONESTROM, U.S. Geological Survey, Menlo Park, California

Staff

KEVIN D. CROWLEY, Study Director
ANGELA R. TAYLOR, Senior Project Assistant

BOARD ON RADIOACTIVE WASTE MANAGEMENT

JOHN F. AHEARNE, *Chair*, Sigma Xi and Duke University, Research Triangle Park, North Carolina
CHARLES MCCOMBIE, *Vice-Chair*, Consultant, Gipf-Oberfrick, Switzerland
ROBERT M. BERNERO, U.S. Nuclear Regulatory Commission (retired), Gaithersburg, Maryland
ROBERT J. BUDNITZ, Future Resources Associates, Inc., Berkeley, California
GREGORY R. CHOPPIN, Florida State University, Tallahassee
RODNEY EWING, University of Michigan, Ann Arbor
JAMES H. JOHNSON, JR., Howard University, Washington, D.C.
ROGER E. KASPERSON, Stockholm Environment Institute, Sweden
NIKOLAY LAVEROV, Russian Academy of Sciences, Moscow
JANE C.S. LONG, Mackay School of Mines, University of Nevada, Reno
ALEXANDER MACLACHLAN, E.I. du Pont de Nemours & Company (retired), Wilmington, Delaware
WILLIAM A. MILLS, Oak Ridge Associated Universities (retired), Olney, Maryland
MARTIN J. STEINDLER, Argonne National Laboratory (retired), Downers Grove, Illinois
ATSUYUKI SUZUKI, University of Tokyo, Japan
JOHN J. TAYLOR, Electric Power Research Institute (retired), Palo Alto, California
VICTORIA J. TSCHINKEL, Landers and Parsons, Tallahassee, Florida

Staff

KEVIN D. CROWLEY, Director
MICAH D. LOWENTHAL, Staff Officer
BARBARA PASTINA, Staff Officer
GREGORY H. SYMMES, Senior Staff Officer
JOHN R. WILEY, Senior Staff Officer
SUSAN B. MOCKLER, Research Associate
TONI GREENLEAF, Administrative Associate
DARLA J. THOMPSON, Senior Project Assistant/Research Assistant
LATRICIA C. BAILEY, Senior Project Assistant
LAURA D. LLANOS, Senior Project Assistant
ANGELA R. TAYLOR, Senior Project Assistant
JAMES YATES, JR., Office Assistant

Preface

This study was undertaken in response to a request to the National Research Council (NRC) from the Department of Energy's (DOE's) Assistant Secretary for Environmental Management. The request was that the NRC conduct a review of the science and technology program designed to address subsurface contamination at the Hanford Site.

The environmental aspects of managing wastes at the Hanford Site have been addressed in several NRC studies by various committees, going back to the mid-1960s. A major focus of these studies has been on the high-level waste storage tanks in the 200 Area, which is located near the center of the site. DOE regards remediation of the 200 Area tank farms as its largest and longest-term environmental challenge. Although much of DOE's past work has focused on the characterization of tank wastes and on the treatment technologies that will be used to stabilize the wastes to make them suitable for disposal, comparatively less effort and attention has been applied to the soil and groundwater at the site.

It has long been known that hazardous wastes have leaked from storage tanks into the underlying soil. In addition, there were direct discharges of liquid waste streams to cribs and ponds on the site when Hanford was producing plutonium for the U.S. nuclear weapons program. There are large uncertainties in the quantities and current locations of materials that were released to the Hanford subsurface. These uncertainties are due in part to the difficulty and expense associated with characterizing soils in the vicinity of the waste tanks and also to the difficulty in characterizing vadose (i.e., unsaturated) zone contamination in general. Unlike groundwater contamination, which tends to form plumes that can be monitored and characterized, vadose zone contamination can follow narrow and variable flow paths that are difficult to detect.

Prior to the mid-1990s, it was generally thought that the sorption capabilities of the soil in the 200 Area would result in limited migration of waste. In particular, it was thought that radioactive cesium would largely be retained in the top several feet of soil. However, this view was challenged when measurements revealed elevated levels of radionuclides deep beneath the tanks and in the groundwater under the tanks. When this was first reported, there was speculation that the measured concentrations at depth were due to inadvertent contamination during drilling. An alternative theory was that there may be fast flow paths in the vadose zone. While the soils in the 200 Area would, based on their average properties, retain cesium and other radionuclides through chemical sorption, these average properties may not determine all potential flow and transport.

vii

In response to the lack of definitive information about the location and mobility of wastes in the subsurface, and with the encouragement of DOE Headquarters personnel, the Hanford Site management established the Groundwater/Vadose Zone Integration Project. The science and technology component of that project is the subject of this report. As indicated in the table of contents, the report addresses the elements of the Integration Project science and technology plan, including the vadose zone, groundwater, Columbia River, remediation and monitoring, risk, and the System Assessment Capability, which is a risk assessment tool in its early stages of development.

As discussed in detail in this report, one of the committee's major conclusions is that there is a great need for better characterization of the subsurface, especially of the vadose zone. Although other work to understand the processes that contribute to the mobility of the wastes and to the modeling of their migration can be useful, such work is of limited value without additional site data. Such characterization data are also needed to test theories about the processes that are important to waste migration so that models of contaminant migration can be refined. A limiting factor in the collection of such site data is cost, which points to the need for more effective and less expensive technologies for characterization.

The committee has been assisted in its efforts by a high level of cooperation and responsiveness from people at the Department of Energy and in the DOE contractor organizations. We especially note the assistance provided by our three main points of contact: Mark Freshley of the Pacific Northwest National Laboratory (PNNL); Michael Graham of Bechtel Hanford, Inc.; and Michael Thompson of the DOE Richland Operations Office. Mark was the committee's main liaison on the science program and handled many requests from the committee for information. We also thank John Zachara of PNNL, who served as a technical guide to the Hanford Site and its associated science and technology projects, and Roy Gephart, who served as a technical guide and provided a very helpful review of Chapter 2.

The committee was also assisted in its efforts by the Hanford representative from the U.S. Environmental Protection Agency, Doug Sherwood; from the Washington State Department of Ecology, Dib Goswami; and from the Oregon Office of Energy, Dirk Dunning. In addition, the committee was kept informed of the activities of the Integration Project Expert Panel (IPEP). Ed Berkey of Concurrent Technologies, Inc., chairman of IPEP, briefed the committee at its first meeting and kept the committee informed through Kevin Crowley, study director for this project. The committee also had the benefit of discussions with the vice-chairman of IPEP, Mike Kavanaugh of Malcolm-Pirnie, Inc.

Information was provided to the committee in presentations and through other means by Harry Boston of the DOE Office of River

Protection, Gerald Boyd of DOE, Office of Environmental Management (EM) (Headquarters), Wade Ballard of the DOE-Richland Operations Office (RL), Mary Harmon of DOE-EM (Headquarters), and Mike Hughes, president of Bechtel Hanford, Inc. Appendix B includes a full list of the people who made presentations to the committee. Priscilla Yamada of PNNL and Virginia Rohay of Bechtel Hanford, Inc. provided logistical support.

Finally, the support provided to the committee by Kevin Crowley was exceptional. Kevin worked with people at DOE and the Hanford contractor organizations to get the right questions asked of the appropriate people, helped arrange the meeting agendas with the right balance of presentations and time for discussion, and kept track of the numerous "loose ends" that the committee generated as it worked to understand the situation at Hanford and the evolving nature of the Vadose Zone/Groundwater Science and Technology Project. The committee faced a steep learning curve, and Kevin helped to identify and define the important issues and activities that the committee needed to focus on, while simultaneously respecting that the committee was to reach its own conclusions. With the help of Angela Taylor, the committee was kept organized and provided with a nearly unlimited supply of reading material. Angela also handled many of the travel and meeting logistics for the committee.

At the time this report went into final review by the National Research Council, a full set of the presentation materials from the committee's meetings at the Hanford site was available on-line at http://www.bhi-erc.com/projects/vadose/peer/nas.htm.

Chris Whipple
Chair

Reviewer Acknowledgments

This report has been reviewed in draft form by individuals chosen for their diverse perspectives and technical expertise, in accordance with procedures approved by the NRC's Report Review Committee. The purpose of this independent review is to provide candid and critical comments that will assist the institution in making its published report as sound as possible and to ensure that the report meets institutional standards for objectivity, evidence, and responsiveness to the study charge. The review comments and draft manuscript remain confidential to protect the integrity of the deliberative process. We wish to thank the following individuals for their review of this report:

Michael D. Annable, University of Florida
Charles C. Coutant, Oak Ridge National Laboratory
Robert J. Naiman, University of Washington
Donald T. Reed, Argonne National Laboratory
John J. Taylor, Electric Power Research Institute, Inc. (retired)
Peter J. Wierenga, University of Arizona
James G. Wenzel, Marine Development Associates, Inc.

Although the reviewers listed above have provided many constructive comments and suggestions, they were not asked to endorse the conclusions or recommendations nor did they see the final draft of the report before its release. The review of this report was overseen by George Hornberger, University of Virginia, appointed by the National Research Council, who was responsible for making certain that an independent examination of this report was carried out in accordance with institutional procedures and that all review comments were carefully considered. Responsibility for the final content of this report rests entirely with the authoring committee and the institution.

Contents

SUMMARY, **1**
1 INTRODUCTION AND TASK, **6**
2 HANFORD SITE BACKGROUND, **11**
3 OVERVIEW OF THE INTEGRATION PROJECT, **40**
4 SYSTEM ASSESSMENT CAPABILITY, **51**
5 INVENTORY TECHNICAL ELEMENT, **66**
6 VADOSE ZONE TECHNICAL ELEMENT, **79**
7 GROUNDWATER TECHNICAL ELEMENT, **100**
8 COLUMBIA RIVER TECHNICAL ELEMENT, **108**
9 MONITORING, REMEDIATION, AND RISK TECHNICAL
    ELEMENTS, **125**
10 IMPROVING S&T PROGRAM EFFECTIVENESS, **141**
REFERENCES, **154**
APPENDIXES
    A BIOGRAPHICAL SKETCHES, **163**
    B INFORMATION-GATHERING MEETINGS, **168**
    C SCALING ISSUES APPLICABLE TO ENVIRONMENTAL
    SYSTEMS, **171**
    D ACRONYMS, **180**

Summary

The Assistant Secretary for Environmental Management at the U.S. Department of Energy (DOE) requested that the National Research Council review the Integration Project's science and technology (S&T) program at the Hanford Site and provide recommendations to improve its technical merit and relevance to DOE's remediation decisions, with particular attention to the following issues:

1. the technical merit of the S&T work to be carried out under the program, including its likely contribution to advancing the state of scientific knowledge;
2. the relevance and timeliness of the planned S&T work to DOE remediation decisions at the Hanford Site; and
3. the potential applicability of S&T results to contamination problems at other DOE sites.

The requested recommendations are provided in this report. The summary is organized according to the three points of the study charge shown above.

CHARGE 1: ASSESS THE TECHNICAL MERIT OF THE S&T WORK TO BE CARRIED OUT UNDER THE PROGRAM, INCLUDING ITS LIKELY CONTRIBUTION TO ADVANCING THE STATE OF SCIENTIFIC KNOWLEDGE

The committee reviewed the S&T projects under way or planned in the seven technical elements of the S&T program (inventory, vadose zone, groundwater, Columbia River, monitoring, remediation, and risk). The committee also reviewed the Integration Project's System Assessment Capability to identify important knowledge gaps that should be addressed by S&T. Detailed comments and recommendations are provided in Chapters 4 through 9 of this report.

The S&T program is at an early stage of development, and many aspects of the program exist only on paper. Moreover, detailed written plans do not exist for most individual S&T projects. Consequently, this review of the S&T program is based primarily on committee members' general knowledge and understanding of relevant scientific and engineering disciplines and Hanford Site problems.

The committee concludes that, in general, the work to be carried out under the S&T program appears on the surface to be technically meritorious and is likely, in at least some cases, to make important contributions to advancing scientific knowledge. This conclusion is qualified, however, because documentation on most projects was insufficient to evaluate in detail either the precise scope of work to be done or its technical merit.[1] Consequently, the committee's evaluations are based on the stated objectives and expected results, not on actual results or detailed study designs or work plans.

Although the S&T program has made a good start, its success is by no means guaranteed. The committee concludes that improvements are needed in the processes used to document and select S&T projects, and to this end, the committee offers the following two recommendations:

1. The Integration Project should develop and implement guidelines for documenting the objectives, technical study designs, work plans, work products, work schedules, and costs for its S&T projects (Chapter 10).

2. Peer review should be used for program prioritization, selection of S&T projects to be funded, and periodic assessments of multiyear projects to ensure that they continue to meet program objectives. Most immediately, peer review should be established to provide continuing oversight of the vadose zone field transport studies (Chapter 6).

CHARGE 2: ASSESS THE RELEVANCE AND TIMELINESS OF THE PLANNED S&T WORK TO DOE REMEDIATION DECISIONS AT THE HANFORD SITE

The committee finds that there is a long-term need for S&T to support cleanup and stewardship of the Hanford Site. According to DOE, environmental cleanup at Hanford is slated to last until at least 2046 and to cost upward of $85 billion, and after this active phase of cleanup is complete the federal government's stewardship responsibilities will last for centuries. The knowledge and technology needed to address the most difficult problems at the site do not yet exist, and advances will not be possible without continuing investments in S&T.

The committee also finds that given the technical and organizational complexity of the task, the Integration Project has made a good start in creating an S&T roadmap, defining and initiating an S&T program, and fulfilling the promise of its mission. The Integration Project appears to have rapidly developed an S&T portfolio that blends well with

---

[1]There were two clear exceptions to this statement: Environmental Management Science Program projects, and some of the projects under the Vadose Zone Technical Element, particularly the vadose zone transport field studies, were well documented.

the activities and needs of the core projects and in this sense is performing well in its "integration" role.

The S&T work under way by the S&T program appears, in general, to be broadly relevant to remediation decisions to be made at the site. This conclusion is qualified, however, because the program lacks a systematic framework for identifying and addressing the uncertainties in knowledge, or "knowledge gaps," that are an impediment to progress in site cleanup. At this early stage of site cleanup, major knowledge gaps are relatively easy to identify, and the S&T program appears to have focused its limited resources on many of these important gaps, particularly with respect to characterization of contamination and hydrologic properties of the vadose zone. However, this ad hoc approach probably will not work as well as the cleanup program matures and a long-term stewardship program is initiated and implemented. In particular, this approach will make it difficult to uncover long-term research needs, which are not easily identified, even in well-planned programs.

The committee's review of the technical elements uncovered several knowledge gaps that are not now being addressed adequately by the S&T program. The committee recommends that addressing these gaps, which are summarized below, should be made a high priority of the S&T program:

- Development of cost-effective strategies and methods for characterization of contaminant distributions and subsurface properties of the vadose zone (Chapter 5).
- Development of advanced monitoring methods for the vadose zone and Columbia River (Chapters 6 and 7).
- Development of improved barrier technologies, including surface barriers, vertical and inclined cutoff barriers, and reactive barriers (Chapter 9).
- Evaluation of the probabilities and consequences of extreme events on Hanford Site contaminants—particularly the effects of catastrophic glacial flooding, which has inundated the site repeatedly during the last 100,000 years and as recently as 15,000 years ago (Chapter 9)—as well as the sensitivity of long-term impacts estimated by the System Assessment Capability to the assumed 1,000-year time scale for peak risk.

The committee also recommends some reprioritization of S&T work to improve its timeliness and relevance to Hanford Site cleanup decisions:

- The planned work on upscaling under the auspices of the Vadose Zone Technical Element should be initiated as soon as possible

(Chapter 6). The lack of early emphasis on an upscaling framework is a serious weakness of the current plans because this framework should play a central role in the design of the field experiments, and also can be used to more directly assess the impact of new information in remediation decisions, thereby providing a basis for setting research priorities.

- Most of the planned work on generic issues under the auspices of the Risk Technical Element should not be funded by the Integration Project (Chapter 9). This work is more appropriate for national research programs in DOE and other federal agencies.

The timeliness of the planned work is difficult to judge given the lack of a strong linkage between S&T and site decisions, as noted previously, as well as the small size and instability of the budget for the S&T program. On the one hand, the small size of the budget ($4.6 million in fiscal year 2001) may not allow substantial progress to be made in addressing the identified knowledge gaps. Moreover, the lack of stable funding is impeding the Integration Project's ability to plan and execute its work—in fact, current S&T work schedules already have been delayed by reductions from the planned budget baseline developed in 1998. On the other hand, S&T is being carried out by other organizations at Hanford and DOE Headquarters, so the total investment in S&T is much greater than indicated by the Integration Project's S&T budget. S&T work across the Office of Environmental Management (EM) and the Hanford Site is not organized or reviewed on a system basis, however, so it is not clear how projects and budgets are being prioritized.

Relevance and timeliness may also be affected by a lack of clarity concerning project ownership. That is, it is not clear who "owns" the S&T projects or whether these owners are being held accountable for progress and costs. Successful management structures usually have clear lines of authority and accountability, and many organizations vest authority and accountability in a single centralized entity. The current structure does not appear to provide this clear management responsibility.

The committee offers the following three recommendations to improve the relevance and timeliness of the S&T work performed under the auspices of the Integration Project:

1. The Integration Project should develop and implement a system for prioritizing its S&T activities to provide the information that Hanford Site management will need to make sound and durable cleanup and stewardship decisions. Procedures are needed to identify key uncertainties in knowledge and determine whether and how these uncertainties could be reduced cost-effectively through further S&T work.

To this end, the System Assessment Capability (Chapter 4) may be a useful tool to help set S&T priorities.

2. The Integration Project should, with the help of EM as necessary, perform a system-based analysis of its funding needs for the S&T program once it develops the prioritization process recommended above. Such an analysis can provide a sound basis for technically defensible funding requests.

3. The Integration Project should review its organization to ensure that ownership, authority, and accountability for the S&T program are clearly defined and assigned. Given the number of organizations involved in S&T and cleanup activities at the Hanford Site, help from DOE management above the level of the Integration Project may be needed to carry out this recommendation.

CHARGE 3: ASSESS THE POTENTIAL APPLICABILITY OF S&T RESULTS TO CONTAMINATION PROBLEMS AT OTHER DOE SITES

The committee did not investigate S&T needs at other DOE sites, so it did not devote much time to addressing this part of its charge. In a sense this question is premature because the S&T program is new and few results are available for transfer. The committee's response to this charge is based primarily on members' knowledge of other DOE sites.

The committee judges that many of the results of S&T work at Hanford are potentially applicable to other DOE sites and, more generally, to other contaminated sites in arid regions. The development of the Integration Project S&T roadmap involved experts from national laboratories and other DOE sites. Perhaps as a consequence, many of the current and planned S&T projects address first-order scientific questions—for example, the development of upscaling techniques (Chapter 6, Appendix C), elucidation of radionuclide and chemical fate and transport in the subsurface (Chapters 6, 7), and evaluation of the impacts of contaminants on biological systems (Chapter 8).

Moreover, some of the planned or committee-recommended S&T work could lead to new technologies that could be applied at Hanford and other DOE sites. Most importantly, the development of techniques for environmental characterization and monitoring, especially in the vadose zone (Chapters 5, 6), and the development of new remediation and containment methods, especially subsurface barriers (Chapter 9), could, with appropriate technology transfer, find widespread application across the DOE complex.

1
Introduction and Task

The Hanford Site was established by the federal government in 1943 as part of the secret wartime effort to produce plutonium for nuclear weapons. The site operated for about four decades and produced roughly two thirds of the 100 metric tons of plutonium in the U.S. inventory. Millions of cubic meters of radioactive and chemically hazardous wastes, the by-product of plutonium production, were stored in tanks and ancillary facilities at the site or disposed or discharged to the subsurface, the atmosphere, or the Columbia River.

In the late 1980s, the primary mission of the Hanford Site changed from plutonium production to environmental restoration. The federal government, through the U.S. Department of Energy (DOE), began to invest human and financial resources to stabilize and, where possible, remediate the legacy of environmental contamination created by the defense mission. During the past few years, this financial investment has exceeded $1 billion annually. DOE, which is responsible for cleanup of the entire weapons complex, estimates that the cleanup program at Hanford will last until at least 2046 and will cost U.S. taxpayers on the order of $85 billion (DOE, 1998e).[1]

Although the "final" condition of the site (i.e., the condition of the site when the cleanup program is complete) has not yet been agreed upon by DOE, its regulators, and other interested parties, work is in progress to stabilize waste and restore the environment so that parts of the site can be released for other uses. After DOE cleanup is completed, however, large areas of subsurface contamination will still remain at the site, including groundwater contamination, and there will be large burial grounds that contain waste from both the defense and the cleanup missions. The cost and duration of the cleanup effort cited above do not account for the long-term investments that will be required to manage these contaminated areas until they no longer pose a hazard to humans or the environment.[2]

One of the most difficult cleanup problems at the Hanford Site involves remediation of the underground high-level waste storage tanks

---

[1] Life-cycle cost estimate fully escalated to year of expenditure. The estimated life-cycle cost in constant fiscal year 1998 dollars is about $51 billion.
[2] The report *Long-Term Institutional Management of U.S. Department of Energy Legacy Waste Sites* (National Research Council, 2000c) discusses these long-term management challenges. See also *From Cleanup to Stewardship* (DOE, 1999f) and *A Report to Congress on Long-Term Stewardship* (DOE, 2001).

Introduction and Task

and the underlying soil and groundwater in the 200 Area (see Chapter 2). There are 177 underground storage tanks at the site, which collectively contain about 54 million gallons of high-level waste generated from plutonium separation processes. According to DOE, 67 of these tanks are known or suspected to have leaked high-level waste into the subsurface, and it is now recognized that some of this leaked material has reached groundwater.

Part of the motivation for this National Research Council (NRC) study grew out of suggestions that the subsurface migration of radionuclides that leaked from these tanks was more extensive than had been predicted (see DOE [1997b] for details). It had been predicted that most radionuclides from these tank leaks would be effectively sorbed onto minerals contained in the subsurface sediments, thereby retarding their migration to groundwater. This assessment appeared to be supported by numerical models developed to predict radionuclide transport beneath the tanks.

Such predictions were called into question, however, by actual measurements of radionuclide (cesium-137) distributions in boreholes around and beneath the tanks in the SX Tank Farm beginning in 1994 (DOE, 1996). These measurements suggested that cesium-137 had migrated greater than 38 meters (125 feet) beneath the SX Tank Farm (DOE, 1998a, pp. 4.50-4.52).[3] Although the actual extent of deep radionuclide migration and the mechanisms for such migration remain unclear (DOE, 1997b), such observations have fueled public concerns and drawn attention to Hanford vadose zone issues in high levels of government (GAO, 1998).

Indeed, this discovery received a great deal of attention by the media and prompted congressional inquiries and a General Accounting Office investigation (GAO, 1998). In response, and with the strong encouragement of DOE Headquarters, Hanford Site management established the Groundwater/Vadose Zone Integration Project in 1997 to coordinate and provide scientific and technical support for waste management and cleanup efforts under way at the site.

The Integration Project was created through a memorandum of understanding among three preexisting organizations at the Hanford Site (see Chapter 3) and is being led by Bechtel Hanford, Inc., with oversight from DOE. The project was established with the following five objectives (DOE, 1998d, p. 1.1):[4]

---

[3]Subsequently, technetium-99 was detected in the groundwater beneath these and other tank farms in the 200 East Area (PNNL, 1999, p. 6.38; DOE, 1998, p. 4-50).
[4]The objectives given here are direct quotes from DOE (1998d). These objectives have been reworded in a subsequent document (DOE, 2000a).

1. Integrate all Hanford Site GW/VZ [groundwater/vadose zone] related work scope.
2. Predict current and future impacts resulting from contaminants that have been (or are predicted to be) released to the soil column at the Hanford Site.
3. Provide a sound science and technology (S&T) basis for site decisions and actions.
4. Promote open and honest involvement of Tribal Nations, regulators, and stakeholders so that project outcomes reflect expressed interests and values.
5. Establish an independent technical peer review.

As discussed in more detail in Chapter 3, the Integration Project is responsible for developing and conducting assessments to determine the effects of chemical and radioactive contaminants on groundwater, the Columbia River, and users of the river's resources. The project is not directly responsible for waste management or cleanup activities at the site. These tasks are the responsibility of the three Hanford Site organizations that signed the memorandum of understanding that gave rise to the Integration Project.

At the request of DOE Headquarters, two technical teams were established to provide peer review of Integration Project activities as called for in the fifth program objective: The Integration Project Expert Panel (IPEP)[5] was created in 1998 to provide advice and recommendations on key programmatic, technical, and administrative issues affecting the success of the Integration Project. This group has been meeting quarterly and has issued several reports that address various aspects of the integration effort at the site.[6] In addition, the Assistant Secretary for Environmental Management requested that the National Research Council review the science and technology S&T program established under the auspices of the Integration Project as called for by the third program objective. That request led to the current study, the results of which are summarized in this report.

---

[5]IPEP consists of eight technical experts: Edgar Berkey, chair, and members Randy Bassett, John Conaway, James Karr, Michael Kavanaugh, John Matuszek, Ralph Patt, and Peter Wierenga.

[6]The expert panel's reports are available on-line at http://www.bhi-erc.com/projects/vadose/peer/ipep.htm.

Introduction and Task 9

SCOPE OF THIS STUDY

The NRC was asked to review the S&T program and to provide recommendations to improve its technical merit and relevance to DOE's remediation decisions, with particular attention to the following issues:

- the technical merit of the S&T work to be carried out under the program, including its likely contribution to advancing the state of scientific knowledge;
- the relevance and timeliness of the planned S&T work to DOE remediation decisions at the Hanford Site; and
- the potential applicability of S&T results to contamination problems at other DOE sites.

The chair of the National Research Council appointed a committee of 14 experts (Appendix A) to undertake this study. The committee met six times to gather information, deliberate on the issues, and develop this report. Three meetings were held in Richland, Washington, near the Hanford Site, so that the committee could receive briefings from DOE staff and site contractors, obtain comments from interested stakeholders, and tour the Hanford Site to see first-hand the cleanup activities and ongoing scientific work. A list of briefings received by the committee at its meetings is provided in Appendix B.

REPORT CONTENT AND ORGANIZATION

The committee's review of the Integration Project's S&T program is organized as follows: Chapters 2 and 3 provide background information on the Hanford Site and the Integration Project. Chapter 4 provides a discussion of the System Assessment Capability, an Integration Project-developed risk assessment tool to estimate quantitative effects of contaminant releases. Chapters 5 through 9 provide reviews of the technical elements of the science and technology program, and Chapter 10 provides programmatic-level recommendations.

The S&T program is at an early stage of development—the draft program plan (DOE, 1998d) was completed in fiscal year 1998, and funding for scientific work was provided beginning in fiscal year 1999. As a result, many aspects of the program exist only on paper, and there is relatively little scientific output on which to judge program effectiveness. In fact, as noted repeatedly in subsequent chapters, detailed written plans in individual S&T projects do not exist for many elements of the program, although there are a few notable exceptions.[7]

Consequently, the reviews of the S&T program elements that are provided in this report are based primarily on committee members' general knowledge and understanding of relevant scientific and engineering disciplines and Hanford Site problems. Except as noted explicitly in the following chapters, none of the committee's comments should be construed as an endorsement of specific individual projects. Rather, the committee's comments address general directions of the S&T program and the apparent appropriateness of program priorities.

The committee has adopted a long-term perspective in its review of this program in recognition of the fact that the DOE clean up program is likely to last for several decades. Even then, there will be a need for continuing management of residual contamination. Consequently, there will be a need for S&T beyond that required to meet near-term milestones and regulatory requirements. Indeed, the S&T work is likely to continue for many years and, if done well, could substantially and positively impact cleanup decisions at the site.

---

[7]Primarily the S&T work under the auspices of the Vadose Zone Technical Element, for which detailed planning documents are available (see Chapter 6).

2
Hanford Site Background

The region around the present-day Hanford Site was occupied by Native Americans for more than 10,000 years before the arrival of the first European American explorers, the Meriwether Lewis and William Clark party, in 1805. Euro-American settlement of the area was promoted by several events: the relinquishment of Indian lands to the government at the Treaty Council of 1855 and military action against Indian resistance in 1858, and the development of irrigation canals and construction of the railroad in the 1880s and 1890s—the latter of which led to the founding of the towns of Kennewick and Pasco. By the early 1940s, the region had a population of about 19,000, supported mostly by farming and ranching.

In December 1942, an officer assigned to the Army Corps of Engineers Manhattan Engineering District and two DuPont engineers began a tour of the western United States to locate a site for a highly classified "atomic" project associated with the war effort. They were seeking a large tract of land with abundant cold water and electricity supplies that was also isolated from highways, railroads, and population centers. After visiting a region along the Columbia River near its confluence with the Yakima River (Figure 2.1), they reported to General Leslie R. Groves, head of the Manhattan Engineering District, that the site "was far more favorable in virtually all respects than any other" (Gerber, 1992). By March 1943, Groves had acquired about 500,000 acres (almost 800 square miles) of land at a cost of a little more than $5 million, and ground was broken for the world's first production facility to make plutonium for atomic weapons (Rhodes, 1986). The site was first designated as Site W and later as the Hanford Engineering Works.

The site design (Figure 2.1) called for three graphite-moderated "atomic piles," or reactors, to be built at 6-mile (about 10-kilometer) intervals along the Columbia River. These areas are referred to collectively as the "100 Areas" and individually by the reactor designation, for example, the "100-B" Area for the B-Reactor. These reactors would irradiate fuel slugs made from natural uranium[1] to create plutonium-239, which had been made in minute quantities for the first time at the Radiation Laboratory (now the E.O. Lawrence Berkeley National Laboratory) in 1941. The river water was needed to cool the piles, which operated at about 200°C. Some 10 to 15 miles (16 to 24 kilometers)

---

[1]The graphite piles were effective neutron moderators and absorbed few neutrons, making it possible to use natural uranium to fuel the reactors. Later, slightly enriched uranium was used to fuel the reactors to increase plutonium production rates.

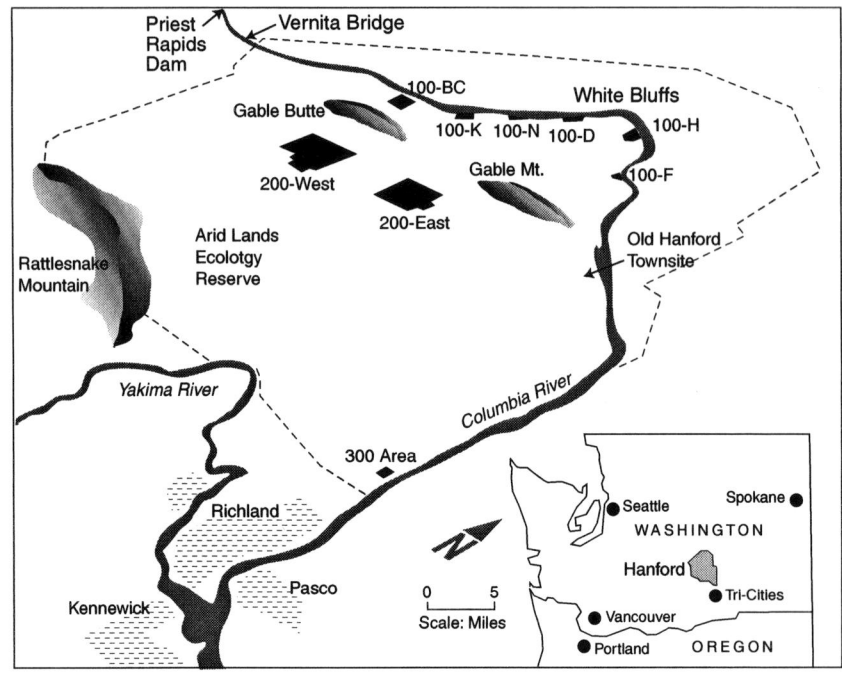

Figure 2.1 Plan view of the present-day Hanford Site showing locations of major plutonium production areas. SOURCE: BHI, 1999, Figure 1-1; DOE, 1998a.

south, on a plateau near the center of the site behind two elevated ridges called Gable Mountain and Gable Butte, two other industrial sites were established, referred to as the 200 East Area and 200 West Area, each containing two massively shielded chemical processing plants to dissolve the irradiated uranium slugs to recover plutonium. The recovered plutonium would then be shipped off-site[2] for processing to make

---

[2]Plutonium was recovered as a nitrate paste, which was shipped to Los Alamos, New Mexico, for conversion to metallic plutonium. A facility to make metallic plutonium at the Hanford Site—the Plutonium Finishing Plant—was constructed in 1949.

plutonium metal that would form the cores of atomic bombs. Sixty-four underground storage tanks were initially constructed near these plants to store the highly radioactive liquid waste from processing operations (Figure 2.2). Additional facilities were constructed downstream of the reactors to manufacture the uranium slugs, in the 300 Area. Given the

Figure 2.2 Construction of single-shell tanks in the BX Tank Farm, 1947. The partially constructed tank in the front-right portion of the photo is filled with liquid, presumably for leak testing. After construction of the steel shells, the tanks were encased in concrete shells and domes, as shown at the left-center and left-rear of the photo. The tanks were constructed below grade (note land surface at the rear of the photo) to provide radiation shielding. SOURCE: David Briggs, Pacific Northwest National Laboratory, Negative 1313.

large potential hazards involved with the operation of these first-of-a-kind facilities, the site design called for these reactors and processing plants to be separated by large distances to minimize potential impacts from routine radionuclide releases as well as catastrophic accidents.

By May 1945, barely two years after groundbreaking, Hanford had produced enough plutonium for the first test of a plutonium bomb, which was carried out at the Trinity Site in New Mexico on July 16. After this successful test, another bomb made from Hanford plutonium (code-named "Fat Man") was dropped on Nagasaki, Japan, on August 9, 1945, thereby forcing an end to the war in the Pacific.[3] By the end of the second world war, the Hanford Site contained more than 500 buildings, almost 400 miles of roadway, and about 160 miles of railroad. A nearby town (Richland) was expanded to house more than 17,000 workers and their families. The total cost of construction was about $230 million (Gerber, 1992).

The Hanford Site was expanded several times after the war to meet national security needs (Table 2.1; see DOE, 1998f, for details). After President Harry Truman's declaration of the Cold War with the Soviet Union in March 1947, Hanford embarked on a $350 million expansion that added two new reactors, a plant to produce metallic plutonium, and new underground high-level waste storage tanks. Following the Soviet Union's detonation of its first atomic bomb in August 1949, a second expansion was undertaken that added yet another reactor, the REDOX chemical processing plant, additional underground waste storage tanks, and two waste evaporators to reduce the large volumes of tank waste being produced from chemical processing operations.[4] The third and final expansion of the Hanford Site occurred during the peak of Cold War tensions during the Eisenhower, Kennedy, and Johnson administrations: three more reactors were built along the Columbia River, another chemical processing plant (PUREX) went into operation, and additional underground waste storage tanks were constructed.

---

[3]An atomic bomb was dropped on Hiroshima, Japan, on August 6, 1945. This bomb, code-named "Little Boy," used uranium-235 as the nuclear explosive. The uranium was produced at the Oak Ridge Site in Tennessee, which was also established during the Manhattan Project.

[4]During this expansion period, many other sites were also established to aid the Cold War effort, most notably the Nevada Test Site, the Idaho Reactor Testing Station (now the Idaho National Engineering and Environmental Laboratory), the Savannah River Site in South Carolina, the Rocky Flats Site in Colorado, the Pantex Plant in Texas, the Fernald Site in Ohio, and the Paducah Plant in Kentucky.

TABLE 2.1 Chronology of Major Production Facilities at the Hanford Site

Facility	Operation Start Date	Operation End Date
Production Reactors		
B-Reactor	1944	1968
D-Reactor	1944	1967
F-Reactor	1945	1965
H-Reactor	1949	1965
DR-Reactor	1950	1964
C-Reactor	1952	1969
KW-Reactor	1954	1970
KE-Reactor	1955	1971
N-Reactor	1963	1987
Fuel Processing Facilities		
T-Plant	1944	1956
B-Plant	1945	1952
REDOX	1952	1967
U-Plant	1952	1958
PUREX	1956	1990
Materials Processing		
Plutonium Finishing Plant	1949	1989
High-Level Waste Tanks		
B-Tank Farm	1945	Inactive
T-Tank Farm	1945	Inactive
C-Tank Farm	1946	Inactive
U-Tank Farm	1946	Inactive
BX-Tank Farm	1948	Inactive
TX-Tank Farm	1949	Inactive
BY-Tank Farm	1950	Inactive
S-Tank Farm	1951	Inactive
TY-Tank Farm	1953	Inactive
SX-Tank Farm	1954	Inactive
A-Tank Farm	1956	Inactive
AX-Tank Farm	1965	Inactive
AY-Tank Farm (D)	1976	Still in service
AZ-Tank Farm (D)	1976	Still in service
SY-Tank Farm (D)	1977	Still in service
AW-Tank Farm (D)	1980	Still in service
AN-Tank Farm (D)	1981	Still in service
AP-Tank Farm (D)	1986	Still in service

Note: The "inactive" tanks contain mostly saltcake, sludge, and some drainable liquids, but they are no longer being used for storage of liquid waste. (D) denotes double containment tank.
SOURCES: DOE, 1998f, Table 2.3.6; tank data from Brevick, 1994, 1995a, 1995b, 1995c.

Production at the Hanford Site began to decline after 1965 in response to decreased national needs for plutonium and other nuclear materials. By 1972, all but one plutonium production reactor was shut down. The last reactor operated until 1987, mainly to produce electricity for the regional power grid.[5]

Twenty-eight additional underground waste storage tanks, each having a storage capacity of between 1.0 million and 1.1 million gallons, were constructed and began receiving waste between 1976 and 1986. These tanks have a double-shell design and are used to hold newly generated waste, as well as waste pumped out of older single-containment tanks, some of which had started leaking in the late 1950s.

At present, all plutonium production reactors and reprocessing plants are permanently shut down. Most facilities have been deactivated, and some are now being torn down. As noted later in this chapter, the Department of Energy (DOE) has also started to remediate contaminated soil and groundwater at the site and to ship transuranic solid waste to the Waste Isolation Pilot Plant (WIPP) in New Mexico.

During its roughly 40 years of operation, Hanford produced about 67 metric tons of plutonium—approximately two-thirds of the nation's plutonium stockpile (DOE, 1998g). In the process, large areas of the site around the production facilities, from the surface to the groundwater, were contaminated with radioactivity and hazardous chemicals. The United States is now spending more than $1 billion per year at Hanford alone to manage residual waste and nuclear materials at the site and to clean up contaminated soil and groundwater, reactors, tanks, chemical processing plants, and ancillary facilities.

WASTE PRODUCTION AND MANAGEMENT

The production of plutonium and other nuclear materials at Hanford consumed more than 95,000 metric tons of uranium fuel and created large volumes of liquid and solid wastes. In the press of the effort to win the Second World War and then to accelerate production during the ensuing Cold War, production of plutonium and other nuclear materials at the site took priority over environmental protection. Most of the high-activity waste produced contains actinides and fission products and is stored in the 200 Area tank farms. In addition, large amounts of radioactive and chemical contaminants were also released into the

---

[5]Eight of the nine reactors at the Hanford site were designed only to produce plutonium. The ninth reactor, designated "N-Reactor," was built with an isolated cooling loop and could produce both plutonium and electricity.

Hanford Site Background

atmosphere, the Columbia River, and the subsurface during the site's 40-year operational history. Until the 1970s, relatively poor records were kept for many of these releases. Some waste continues to be released to the environment today from waste management and cleanup operations at the site. These controlled environmental releases are now regulated by the Environmental Protection Agency and Washington State.

Although plutonium production took priority at the site, there was a concern about potential environmental impacts even from the earliest days of site operations. Programs were established to monitor and limit worker exposures and make environmental measurements of the Columbia River and its aquatic life, site vegetation, wildlife, and groundwater. Extensive studies of the Columbia River ecosystem concentrated on both radionuclide and thermal (heat) releases (e.g., Vaughan and Hebling, 1975; Becker, 1990). After the war, additional studies were made of site sediments to determine their capacity to retard the migration of radionuclides, which were being released into the subsurface along with large volumes of water. As noted elsewhere in this section, some operational practices were modified to reduce waste releases based on these monitoring programs.

Because of incomplete record keeping, an exact mass balance of historical releases of radioactivity and chemicals to the environment at the site does not exist. The Integration Project has established a program to obtain such an estimate, as described in Chapter 5 of this report. The following sections summarize what is currently known about contaminant releases at the site, organized by environmental medium as illustrated in Figure 2.3. More detailed discussions of waste releases can be found on the Hanford web site; see especially the *History of the Plutonium Production Facilities at the Hanford Site Historic District* (DOE, 1997a; http://www. hanford.gov/docs/rl-97-1047/index.htm) and many of the references cited therein.

Releases to the Atmosphere

The operation of production reactors resulted in the release of about 12 million curies of volatile fission products to the atmosphere (Heeb, 1994). Volatile radioisotopes were also released during chemical processing of the fuel to recover plutonium,[6] especially during the war years (Napier, 1992). Emissions from the chemical processing plants were reduced after the war through the use of scrubbers and filters and

---

[6]The chemical processing plants had 200-foot-high vent stacks to disperse these releases.

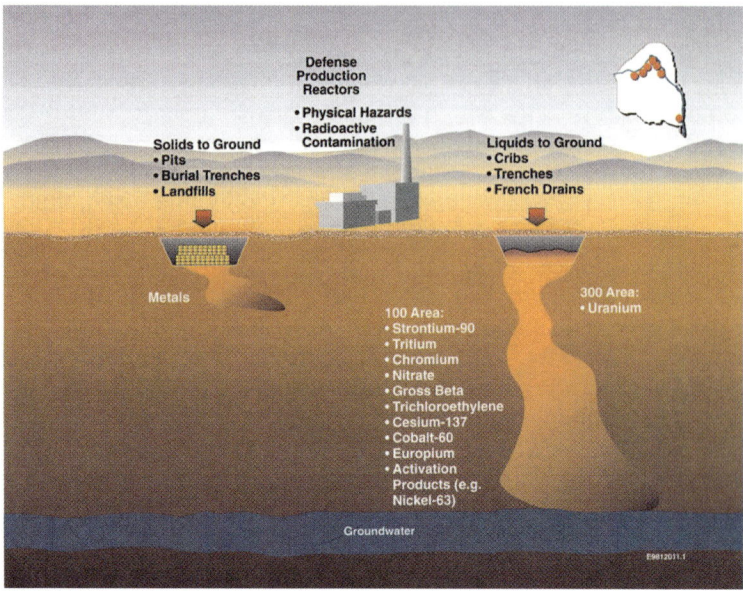

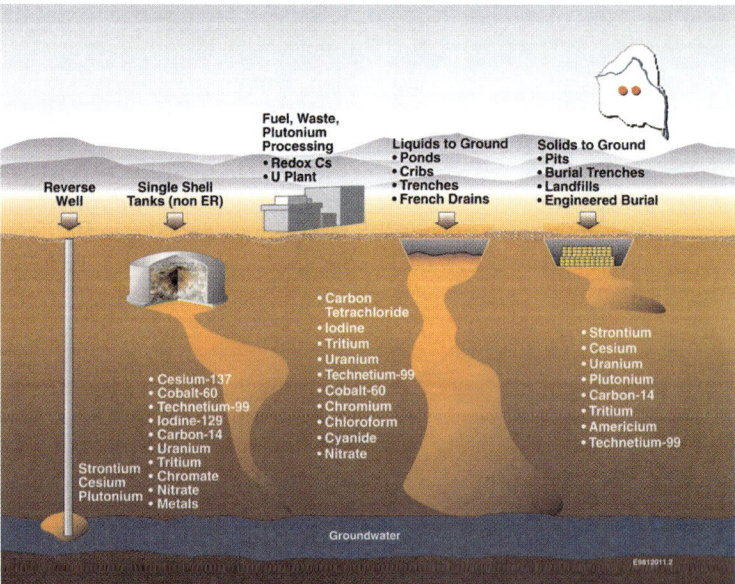

Figure 2.3 Major contaminants and subsurface release pathways in the (A) 100 and 300 Areas and (B) 200 Area at the Hanford Site. SOURCE: DOE, 1998a, Figures 1-5 and 1-6.

Hanford Site Background

by allowing more time for the fuel to "cool" after irradiation to allow short-lived radionuclides to decay.

Once released into the atmosphere, radionuclides were dispersed by atmospheric mixing. The impacts of atmospheric releases of radionuclides at Hanford on human health have been assessed in the Hanford Environmental Dose Reconstruction Project and the Hanford Thyroid Disease Study (National Research Council, 1994a, 1995, 2000b). These studies have shown that iodine-131 (half-life~8 days) contributed most of the radiation dose received by members of the public from atmospheric releases at Hanford.

Releases to the Ground

The release of radionuclides and hazardous chemicals to the ground at the Hanford Site occurred at all of the major production areas. These contaminants are among the most significant potential environmental hazards that exist at the site today—in addition to the spent nuclear fuel, high-level waste, and other nuclear materials under active management at Hanford. These releases can, for convenience, be grouped into the following three categories: (1) solid waste disposal, (2) liquid waste disposal, and (3) accidental releases and discharges.

Solid Waste Disposal

Radioactive and chemically contaminated solid waste has been disposed of in shallow land burial grounds around all of Hanford's production facilities. Almost 70 burial sites containing more than 650,000 cubic meters of waste are known to exist (DOE, 1997d, 2000h). Solid waste was placed in unlined trenches, lined excavations, and underground vaults and consisted of a wide variety of materials, including failed hardware, construction and demolition waste, soil contaminated by spills and leaks, contaminated clothing, and various kinds of process waste.

During the first two decades of site operation, burial grounds were built in close proximity to production facilities, and both chemical and radioactive wastes were disposed with little or no segregation. Moreover, no detailed records were kept of the kinds or amounts of waste disposed. By the 1960s, the burial grounds were centralized, mostly in the 200 Area, and waste segregation and better record-keeping practices were implemented. By the 1970s, all radioactive solid waste was being disposed of in the 200 Area, and transuranic waste was being segregated and stored.[7] Additionally, computerized databases began to be used to track inventories of waste disposed of in the burial grounds.[8] In 1995, the Environmental Restoration Disposal Facility (ERDF)[9] was established between the 200 East Area and the 200 West Area (see Figure 2.1). It now receives most of the solid radioactive and mixed waste[10] generated by cleanup and waste management activities at the site.

The burial grounds in the 200 Area also hold waste generated off-site by other DOE sites and laboratories, universities, the military, and private companies. Most notable, perhaps, is the burial ground in the 200 East Area that holds more than 80 reactor compartments from decommissioned U.S. nuclear submarines. A private-sector organization (U.S. Ecology) also operates a commercial low-level waste disposal facility on land owned by Washington State.

Liquid Waste Disposal

Liquid radioactive and chemical wastes were discharged to the ground at all operating facilities on the Hanford Site. In terms of volume and toxicity, the most significant releases occurred in the 200 Area from chemical processing operations. After irradiation, the fuel was brought to the 200 Area by train, where it was dissolved and chemically processed to recover plutonium, uranium, and sometimes neptunium.[11] These processing operations produced 26 distinct waste streams containing actinides and fission products and a wide range of chemicals, including nitric acid, bismuth phosphate, potassium permanganate, methyl isobutyl ketone, aluminum nitrate, tributyl phosphate, kerosene, ammonium fluoride, and sodium hydroxide.

---

[7]A 1970 Atomic Energy Commission directive required the segregation of transuranic waste and also required that it be placed in retrievable storage. That stored waste is now being shipped to the WIPP in New Mexico for disposal.

[8]The database is now referred to as the Solid Waste Tracking System (see Chapter 5).

[9]The ERDF is a Resource Conservation and Recovery Act- and Comprehensive Environmental Response, Compensation, and Liability Act-compliant land disposal facility for disposal of waste from Hanford cleanup operations. It comprises a series of disposal cells, each measuring about 500 feet on a side and 70 feet deep, with a combined capacity of almost 12 million cubic yards.

[10]Mixed waste contains both radionuclides and hazardous chemicals.

[11]Neptunium was used to make plutonium-238 for radioisotope thermoelectric generators (also known as RTGs) for space applications.

Hanford Site Background

Table 2.2 provides an inventory of the high-level waste produced by chemical processing operations between 1944 and 1988.[12] The numbers given in Table 2.2 are rough estimates only[13] and are based on process knowledge supplemented with records where available. Detailed records of soil discharges were not kept, and even the current high-level waste inventory of specific radionuclides and chemicals in the tanks is not well known.[14] As noted previously, an effort is under way at the site to obtain better estimates of historical releases to aid the long-term cleanup effort.

The following discussion is based on the inventory estimates given in Table 2.2. Chemical processing operations generated more than 500 million gallons of high-level waste with a radionuclide content of about 800 million curies.[15] This waste was transferred to the waste storage tanks by underground transfer lines. Once in the tanks, the waste was subjected to additional treatment to reduce its volume by more than 90 percent, to the 54 million gallons that exist in the tanks today (Table 2.2). This was done using the following processes:

1. Beginning in about 1948, when tank space was in short supply, gravity separation of the solid and liquid fractions of the high-level waste was accomplished using multiple tanks connected in series. Waste was introduced into the upstream tank, and as it cascaded through successive tanks, the solid fraction, which contained most of the actinide elements and strontium, would settle out, leaving a liquid *supernate* that contained cesium and other soluble fission products such as technetium. At the end of the cascade, the supernate was discharged to soil.

2. After cascading was discontinued in the 1950s, the supernate in some tanks was treated with potassium ferrocyanide to precipitate cesium. Once the cesium was removed, the remaining liquid was discharged to soil.

---

[12]The committee is indebted to Roy Gephart, Pacific Northwest National Laboratory (PNNL), who provided some of the background material used in this section and in Table 2.2 and who reviewed a draft of this chapter.

[13]The committee cannot evaluate the accuracy of the estimates given in Table 2.2 but believes that they are likely to be highly uncertain. The numerical ranges shown for some entries in the table represent differences in estimating procedures and do not necessarily represent the uncertainty ranges of the estimates themselves, which have not been determined, in part because the quality of the estimates is unknown.

[14]The waste tanks are highly heterogeneous, and not all of the tanks have been sampled.

[15]This estimate is based on a rough calculation and was provided by Roy Gephart (PNNL).

TABLE 2.2 Inventory of High-Level Waste at the Hanford Site

	Waste Volume (million gallons)	Curies to Ground (millions)[a]	Curies in Facilities (millions)[a]	References
HLW generated	530	—	—	Agnew (1997)
Discharges to soil[b]	120-130	0.065-4.7[c]	—	Waite (1991); Agnew (1997)
Tank Leaks to soil[d]	0.75-1.5	0.45-1.8	—	Waite (1991); ERDA (1975); Agnew (1997)
Evaporator condensates discharged to soil	280	0.003	—	Agnew (1997); Hanlon (2000); Wodrich (1991)
Cooling and processing water	400,000	Negligible	—	DOE (1992a, 1992b)
Cesium and strontium capsules	—	—	140	Final Tank Waste Remediaiton System EIS, Appendix A, Table A.2.2.1
Tank waste	54	—	210-220	Waite (1991); Agnew (1997); Hanlon (2000)
Facilities	—	—	10[e]	Gephart (1999)
Total		0.22-6.5	360-370	

Note: Numbers are rounded to two significant digits from the values given in the references. The numerical ranges represent differences in estimating procedures and do not necessarily represent uncertainty ranges of the estimates themselves, which have not been determined, in part because the quality of the estimates is unknown.
[a]Quantities are decay corrected to the mid to late 1990s.
[b]After cascading through multiple tanks or after chemical treatment to remove cesium.
[c]The lower estimate is for cesium-137 and minor amounts of strontium-90 only.
[d]Estimate does not include leaks from transfer lines and valves.
[e]Radionuclides estimated to remain in plutonium production reactors and chemical separations facilities.

3. The REDOX and PUREX processing plants, which began operations in 1952 and 1956, respectively, produced "self-boiling" waste from increased radioactive decay heating per unit volume being processed.[16] The vapor produced in these boiling tanks was captured, condensed, and discharged to soil.

4. Tank waste was also sent to evaporators, where it was heated to boil off excess water. The resulting vapor was captured, condensed, and discharged to soil, as were the evaporator "bottoms."[17]

The first three processes described above produced over 100 million gallons of mostly low-activity waste, containing approximately 5 million curies of radioactivity that was discharged to soil (second row in Table 2.2). Because it resulted from a distillation process, the evaporator condensate described by the fourth process comprised a large volume of liquid (about 300 million gallons) but only a few thousand curies of radioactivity (fourth row in Table 2.2).

These discharges to soil occurred through a variety of means. Initially, waste was discharged to artificial ponds on the surface of the 200 Area (Figure 2.4). This practice was soon abandoned because of surface contamination problems. Later, injection wells, euphemistically referred to as "reverse wells," were employed, but these too were abandoned after about two years because of plugging problems and concerns that the wells could provide pathways to groundwater. Finally, shallow drainage structures (French drains and cribs) were built starting in 1947 to handle the large volumes of waste (Figure 2.5).

In total, more than 400 million gallons of mostly low-activity and mixed waste were discharged to the subsurface in the 200 Area during operation of the production facilities (Table 2.2). In addition, more than 400 billion gallons of cooling and processing water were also discharged to the ground in the 200 Area during operation of the production facilities (DOE, 1992a, 1992b). These water discharges raised local groundwater levels beneath the 200 Area, thereby creating large groundwater "mounds" that changed local hydraulic gradients and promoted the movement of groundwater contaminants toward the Columbia River (Figure 2.6). Today, these "mounds" stand between 25 and 85 feet above

---

[16] The PUREX plant produced more concentrated high-level waste (between 50 and 1,400 gallons per ton of uranium fuel) than the REDOX (1,000-4,600 gallons per ton) or bismuth phosphate process (2,000 to 25,000 gallons per ton) plants (Agnew, 1997). Consequently, radionuclide concentrations were much higher in the PUREX waste stream.

[17] Evaporator bottoms consist of sediments left over in the evaporator after the waste slurry is removed. Some of these sediments had high levels of radionuclides such as technetium.

Figure 2.4 Low-angle oblique photo of the Gable Mountain pond at the Hanford Site. This pond was closed in 1984 and is now covered with gravel. SOURCE: David Briggs, Pacific Northwest National Laboratory, Negative 62440-23.

Figure 2.5 Construction of a drainage crib at the Hanford Site in 1944. The sawhorse in the excavation provides scale. SOURCE: David Briggs, Pacific Northwest National Laboratory, Negative 3543.

the groundwater levels that existed before the site was established in 1943 (PNNL, 1999). Mound heights are slowly decreasing.

In addition to radionuclides, hazardous chemical waste was also discharged to the subsurface at the site. For example, carbon tetrachloride was discharged in large quantities from the Plutonium Finishing Plant in the 200 Area. Rohay (2000) estimates that approximately 600,000-900,000 kilograms of carbon tetrachloride containing about 200 kilograms of plutonium were discharged to three cribs from 1955 to 1973. This has created a large groundwater plume that extends from these cribs under the Plutonium Finishing Plant, with peak concentrations of carbon tetrachloride that exceed 6,000 micrograms per liter (Figure 2.7).

DOE installed a vapor extraction system and a pump-and-treat system to recover carbon tetrachloride from the vadose zone and groundwater in this area. The soil vapor extraction system operated from 1992 to 1999, and DOE estimates that it recovered about 10 percent of the carbon tetrachloride from the vadose zone. Less than one percent was removed from the groundwater through pump-and-treat operations, which continue through the present. About 24 percent is estimated to have been lost to the atmosphere or biodegraded, and about 65 percent of the carbon tetrachloride is estimated to remain in the subsurface (DOE, 2000e), in either or both the vadose zone and groundwater.

Liquid discharges to the ground occurred in the 100 and 300 Areas of the site as well, but these were not as extensive as releases in the 200 Area. In the 100 Area, both liquid cooling water contaminated with chromium (a corrosion inhibitor) and fission products from broken fuel elements were discharged to trenches near the reactors. Contaminated primary cooling water from the N-Reactor, which had a two-loop cooling system, was also discharged into trenches. In the 300 Area, liquid wastes were collected in a process pond, where they were allowed to percolate into the soil.

According to DOE, groundwater under more than 100 square miles (260 square kilometers) of the site is contaminated above drinking water standards with radionuclides and chemicals discharged to ground in the 100, 200, and 300 Areas. The contaminants include tritium, strontium-90, technetium-99, iodine-129, uranium, carbon tetrachloride, and chromium. Large bodies of contaminated groundwater (groundwater plumes) are flowing southeast and northwest toward the Columbia River at rates of up to several tens of meters per year (Figure 2.8). At present, several plumes release contaminants into the Columbia River. The migration of contaminants is largely controlled by the subsurface characteristics of the site (Sidebar 2.1).

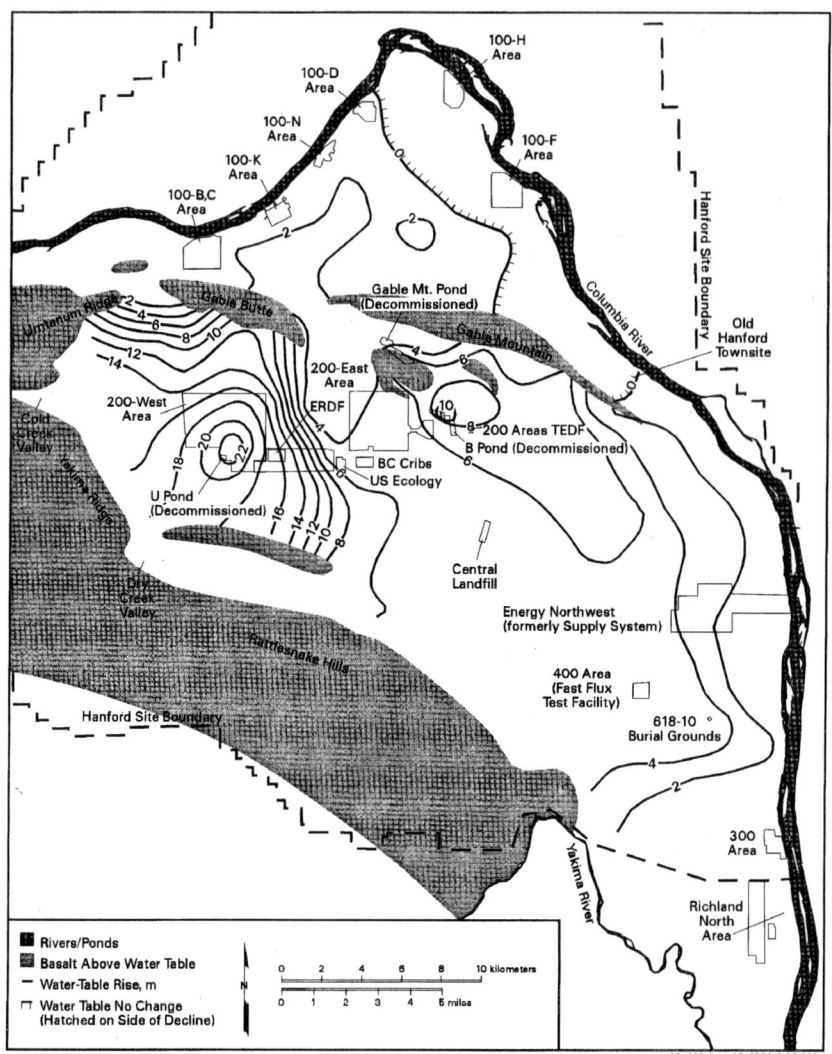

Figure 2.6 Change of groundwater table elevations at the Hanford Site between 1949 and 1979. Contours show the groundwater table rise over this period in meters. SOURCE: PNNL, 1999, Figure 6.1.7.

Hanford Site Background

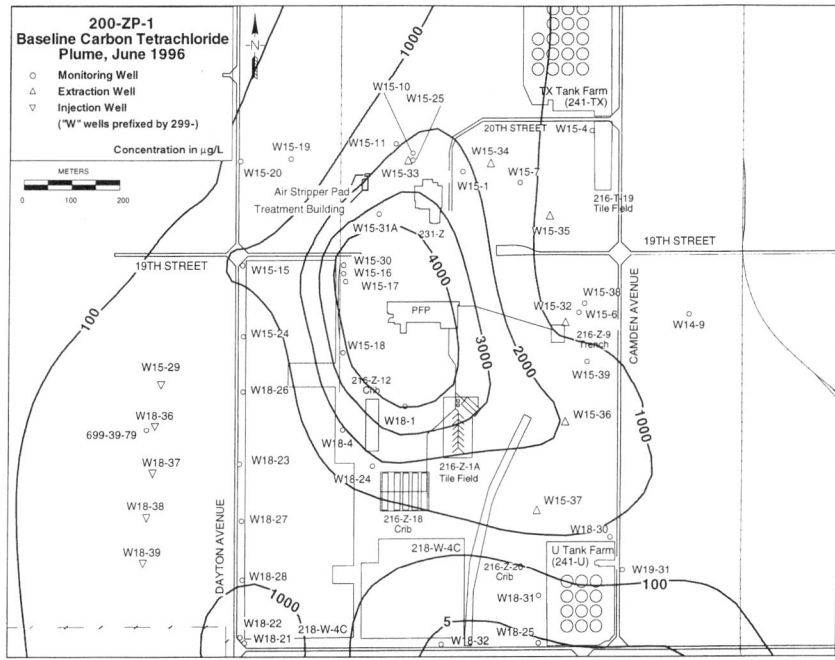

Figure 2.7 Map showing carbon tetrachloride groundwater plume concentration isopleths (in micrograms per liter) in the 200 West Area at the Hanford Site in June 1996. The building marked "PFP" is the Plutonium Finishing Plant. SOURCE: DOE, 2000e, Figure 3-8.

Accidental Releases and Discharges

Accidental releases of chemical and radioactive contaminants occurred at many facilities at the Hanford Site. In the 100 Area, for example, leaking reactor cooling-water retention basins raised local groundwater levels, resulting in the creation of local springs along the riverbank as well as local changes in groundwater flow directions.

The major releases have occurred in the 200 Area and involve

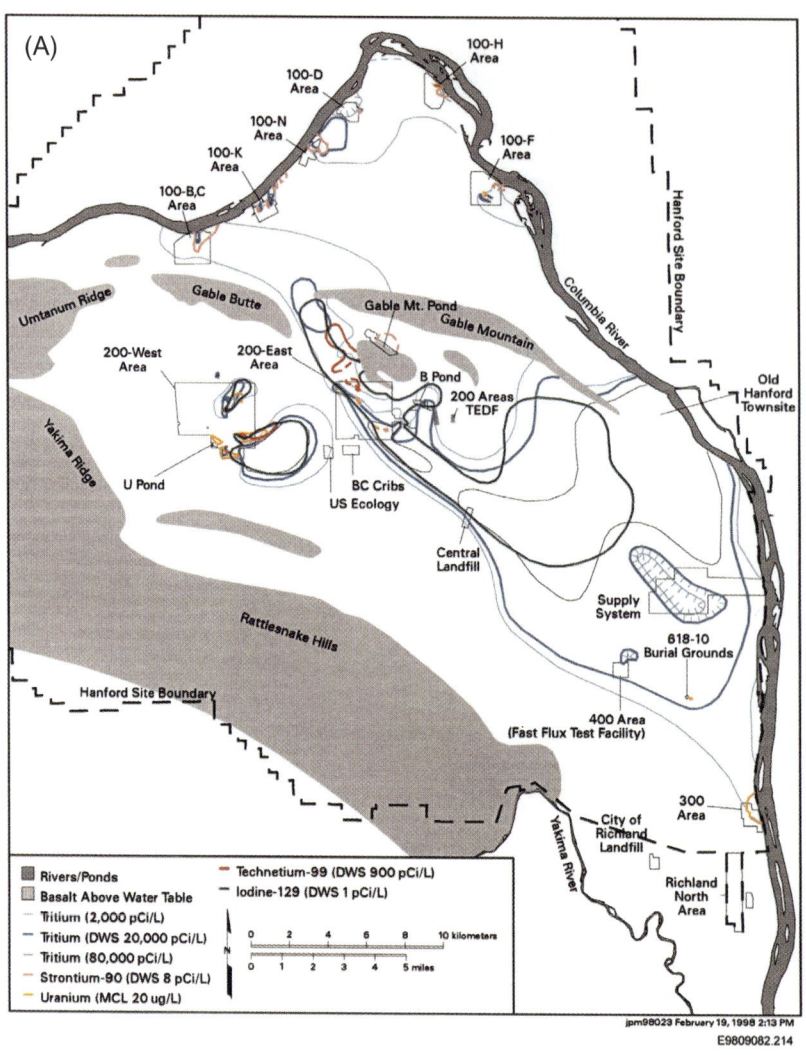

Figure 2.8 Boundaries of major groundwater plumes at the Hanford Site: (A) radionuclide plumes, (B) chemical plumes. SOURCE: DOE, 1998a, Figures 1-3 and 1-4.

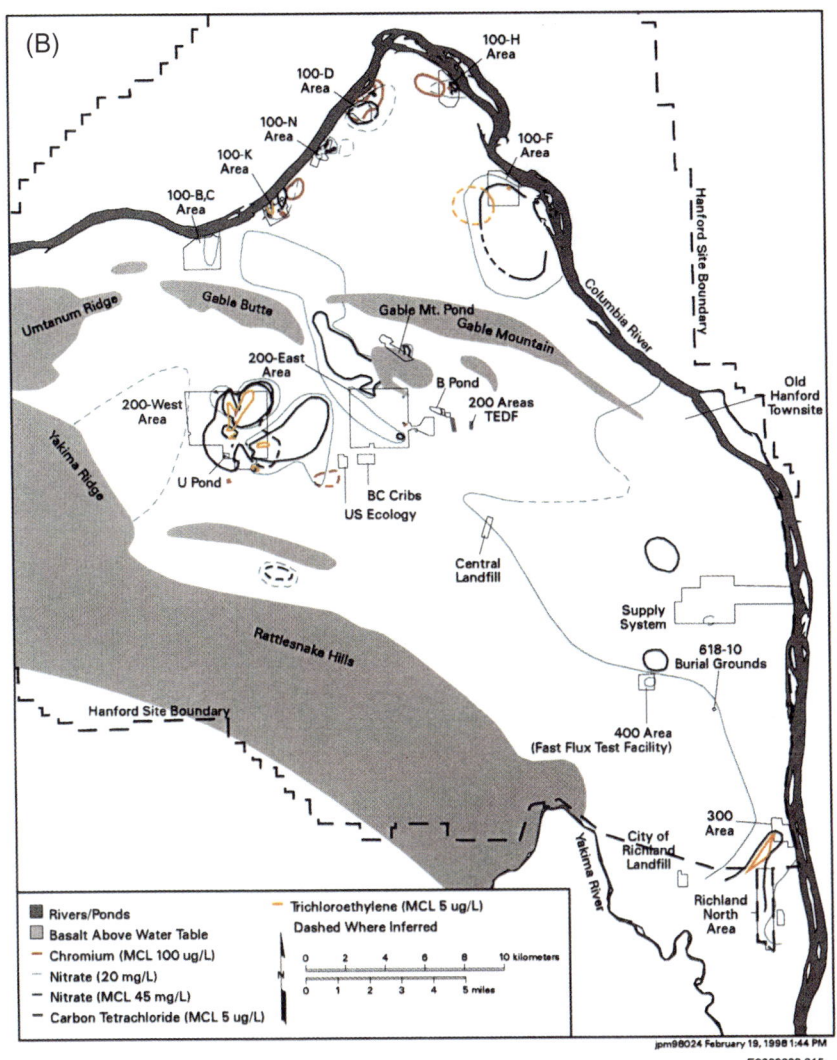

SIDEBAR 2.1 Subsurface Characteristics of the Hanford Site

The Hanford Site is located on the Columbia Plateau, a broad volcanic plain that stretches between the Rocky Mountains and Cascade Range and that is underlain by flood basalts and a variety of fluvial and lacustrine deposits. The flood basalts are part of the Columbia River Basalt Group, which were erupted between about 17 and 6 million years ago over what is now southeastern Washington and northern Oregon. The basalts are interbedded with sedimentary rocks of the Ellensburg Formation, and are overlain by approximately 3 to 8 million-year-old sediments of the Ringold Formation and < 1 million-year-old cataclysmic flood deposits of the Hanford Formation. A generalized east-west section through the site is shown in Figure 2.9.

The surface and near-surface characteristics of the site have been defined by repeated catastrophic flooding over the past 100,000 years (see Sidebar 9.1). The Hanford Site is, in essence, a large, mostly dry riverbed, with interwoven layers of gravel, sand, and silt surrounding occasional bedrock "islands" (e.g., Gable Butte and Gable Mountain). These sediments comprise the Hanford Formation, the thickest accumulations of which occur beneath the site's Central Plateau. The Ringold Formation crops out along White Bluffs (see Figure 2.1) on the northern and eastern shore of the Columbia River and, although not shown on the figure, the Columbia River Basalt Group outcrops on Gable Mountain and Gable Butte.

The subsurface migration of contaminants is controlled to a great extent by the physical and chemical characteristics of the underlying geology at the site. For example, the Hanford Formation contains basalt-rich sediments that are highly sorptive of radionuclides like cesium and strontium. As a consequence, most of these radionuclides have been trapped in the vadose zone. However, the Hanford Formation is also unconsolidated and highly permeable to contaminants, such as tritium and technetium, that are not readily sorbed. This explains why tritium discharged into the subsurface in the 200 East Area has formed large groundwater plumes that extend to the Columbia River (see Figure 2.8a). Such plumes have not formed in the 200 West Area, probably because the groundwater table is located in the more highly consolidated and therefore less permeable Ringold Formation.

Groundwater movement in the highly permeable sediments underlying the site is generally toward the Columbia River, driven by the topography of the regional groundwater table. This movement is modified locally by groundwater barriers (e.g., Gable Butte and Gable Mountain) and by recharge from site operations (see, e.g., Figure 2.0). Of significance to the Columbia River ecosystem is the localized discharge of contaminated groundwater into the bed of the Columbia River, possibly creating zones of elevated contaminant concentrations.

Hanford Site Background

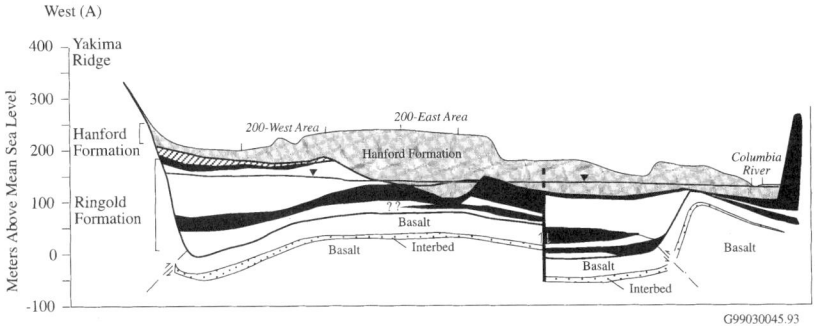

Figure 2.9 Generalized east-west section through the Hanford Site showing the principal geologic formations. SOURCE: PNNL, 1999, Figure 6.1.3.

leaks of high-level waste from tanks and waste transfer lines. Leakage of high-level waste to the subsurface is suspected to have occurred in at least 67 of the 149 single-containment underground waste tanks in the 200 Area (Gephart and Lundgren,1998). The word "suspected" is used to describe these leaks because the single-containment tanks were not designed with systems to detect leaks. Rather, leakage has been inferred by monitoring liquid levels in the tanks and by radiation monitoring in about 800 dry wells[18] drilled in many of the tank farms (Figure 2.10).

The single-containment tanks were constructed beginning in 1945 and had a 20-year design life. The first tank leak, estimated to be around 10,000 gallons, is believed to have occurred in the U Tank Farm in 1956, about 10 years after it was constructed (Table 2.1). The largest leak, estimated to be more than 100,000 gallons, is believed to have occurred in the T-Tank Farm in 1973 (DOE, 1997a). The total amount of leakage from all 67 tanks is estimated to have been between 750,000 and 1.5 million gallons of high-level waste with an activity between about 450,000

---
[18]Wells completed in the vadose zone above the water table.

and 1.8 million curies (Table 2.2). Most of the liquids contained in these leaking tanks have been pumped into double-containment tanks (Gephart and Lundgren, 1998). In some cases the remaining liquids were absorbed by adding diatomaceous earth.

The subsurface in the 200 Area was also been contaminated with uranium from operation of the U-Plant from 1952 to 1958. An estimated

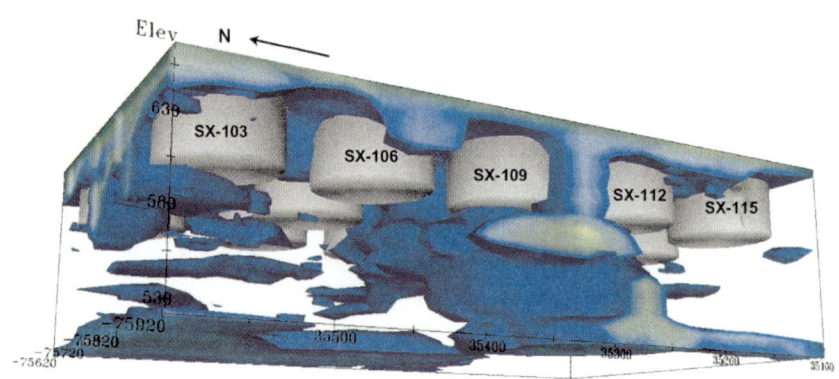

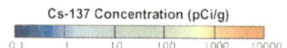

Figure 2.10 Calculated cesium-137 distributions in soil beneath the SX Tank Farm. Vertical lines represent dry wells in which gamma-ray measurements were made to determine cesium concentrations. Tanks labeled in red font are known leakers. SOURCE: DOE, 1998a.

4,000 kilograms of uranium was disposed in two cribs during operation of this plant. Some of the uranium was later remobilized and transported to groundwater beneath the 200 Area when acid waste was inadvertently disposed to these cribs and additional disposal cribs were put into operation nearby. The acid remobilized the uranium in the crib and underlying sediment, and the liquids from a nearby crib transported this remobilized uranium to groundwater. This groundwater contamination is being contained through pump-and-treat operations (DOE, 2000e).

Releases to the Columbia River

There were many releases of radioactive and chemical contaminants to the Columbia River during operation of the production facilities at the Hanford Site, and some releases from contaminated groundwater continue to the present. By far the largest releases occurred from the eight "single-pass" production reactors in the 100 Area, which released about 110 million curies to the river (Heeb and Bates, 1994).[19] Up to 200,000 gallons per minute of treated river water was used to cool these eight reactors, and as the treated water passed through the reactor cores, naturally occurring elements in the water became activated by capturing neutrons. Additionally, a small percentage of the radionuclides released to the water were fission products from damaged fuel elements. The principal contaminants in the reactor effluents are shown in Table 2.3.

Reactor operations also resulted in discharge of liquids into the subsurface around reactor sites, which later migrated through the groundwater and into the river. As noted previously, cooling water contaminated with radionuclides from damaged fuel elements was sometimes diverted into trenches, as was contaminated water from the primary cooling loop on the N-Reactor. Process waste and water treatment chemicals (e.g., sodium dichromate) leaked or were disposed of at the reactor sites. Some of these contaminants continue to leak into the river. Pump-and-treat facilities and other treatment approaches[20] are being implemented to reduce the inflow of these contaminants to the river.

---

[19] As noted elsewhere in this chapter, all of the production reactors except for the N-Reactor were cooled by pumping treated river water directly through the cores. On exiting the cores, the water was held in a retention basin for a few hours before being pumped back into the river. The N-Reactor had a closed primary loop to cool the core. Cooling water from the river was provided in a secondary loop that was isolated from the reactor core.

[20] For example, the oxidation state of chromium is being manipulated in groundwater near the D-Reactor to immobilize it in place and limit its migration into the river.

TABLE 2.3 Selected Radionuclide Releases to the Columbia River from Single-Pass Hanford Reactors, 1944-1971

Radionuclide[a]	Half-Life	Total Curies (millions)
Sodium-24	15 hours	12.6
Phosphorus-32	14.3 days	0.23
Zinc-65	245 days	0.49
Arsenic-76	26.3 hours	2.5
Neptunium-239	2.4 days	6.3

[a] According to the Hanford Dose Reconstruction Project (Farris et al., 1994), these five radionuclides contributed more than 94 percent of the total dose to representative individuals who used Columbia River resources.
SOURCE: Heeb and Bates, 1994.

Production activities in the 200 East Area have created large groundwater contaminant plumes that are discharging nitrate and tritium into the Columbia River downstream of the 100 Area (Figure 2.8). About 3,000 curies, on average, of tritium is discharged into the river each year from the site, based on sampling data (e.g., PNNL, 1999, 2000a) from the river near the upstream and downstream boundaries of the site. The Hanford Site contribution increases the radionuclide load in the Columbia River by about one-third. The remaining radioactivity in the river is from natural or man-made[21] sources upstream of the Hanford Site.

CLEANUP OF THE HANFORD SITE

Hanford Site's defense mission waned in the late 1980s, prompted by the shutdown of the N-Reactor in response to the Chernobyl accident and a thaw in the Cold War, and the focus of site activities shifted from plutonium production to environmental restoration. In 1989, DOE, the State of Washington Department of Ecology, and the U.S. Environmental Protection Agency entered into the Hanford Federal Facility Agreement and Consent Order, also known as the "Tri-Party Agreement," for achieving compliance with CERCLA (the Comprehensive Environmental Response, Compensation, and Liability Act) and RCRA (the Resource Conservation and Recovery Act) provisions of federal

---

[21] Primarily from fallout left over from atmospheric tests of nuclear weapons.

statutes, as well as state environmental protection laws.[22] The Tri-Party Agreement defines and ranks cleanup and waste management commitments, establishes cleanup responsibilities, and provides enforceable milestones for achieving these commitments. Cleanup work at the Hanford Site has proceeded under this agreement since it was signed, although DOE has had to renegotiate many of the agreed-to milestones.

In 1999, DOE released the *Final Hanford Comprehensive Land-Use Plan Environmental Impact Statement* (DOE, 1999a),[23] which lays out its preferred future land use at the Hanford Site after the cleanup program is completed. DOE's preferred alternative (Figure 2.11) includes the following provisions:

- The land surrounding the core of the Hanford Site (the Wahluke Slope north of the Columbia River and Arid Lands Ecology Reserve southwest of the Central Plateau) and Rattlesnake Mountain and Gable Butte will be preserved from impacts from intensive land-disturbing activities (e.g., mining or extraction of nonrenewable resources).
- The Columbia River corridor will have a variety of land uses. The river islands and a quarter-mile buffer zone on each side of the river channel will be preserved to protect cultural and ecological resources. However, the "cocooned" reactors will not be moved for at least 50 years, and remediation will continue as necessary along the river. Additionally, B-Reactor will become a museum. Several sites along the river will be designated for recreational use.
- Most of the Hanford Site will be designated as conservation zones to protect cultural, ecological, and natural resources. However, excavation will be permitted to obtain materials needed for DOE missions—for example, to construct barriers and caps to retard future contaminant movement at waste disposal sites.
- The Central Plateau will be designated as industrial-exclusive use, which would allow current waste management activities to continue and new compatible facilities to be developed.
- The portion of the site north of Richland will be designated as industrial, which would support future DOE missions or commercial industrial development.
- An area in the southeastern portion of the site will be designated for research and development to support DOE's continuing

---

[22]CERCLA provisions govern the cleanup of contaminated sites, whereas RCRA provisions govern the treatment, storage, and disposal of waste generated at the site.
[23]Available on the Hanford Website at http://www.hanford.gov/eis/hraeis/hraeis.htm.

energy research mission. This area now contains the Laser Interferometer Gravitational Observatory.

DOE recognizes that cleanup is likely to be incomplete, even in the areas designated for recreation and preservation, and that deed restrictions and continuing (in some cases, perpetual) institutional management will be required over much of the site to protect public and environmental health.

Within the Central Plateau, the 200 Area will serve the site's continuing waste management mission. The major waste management

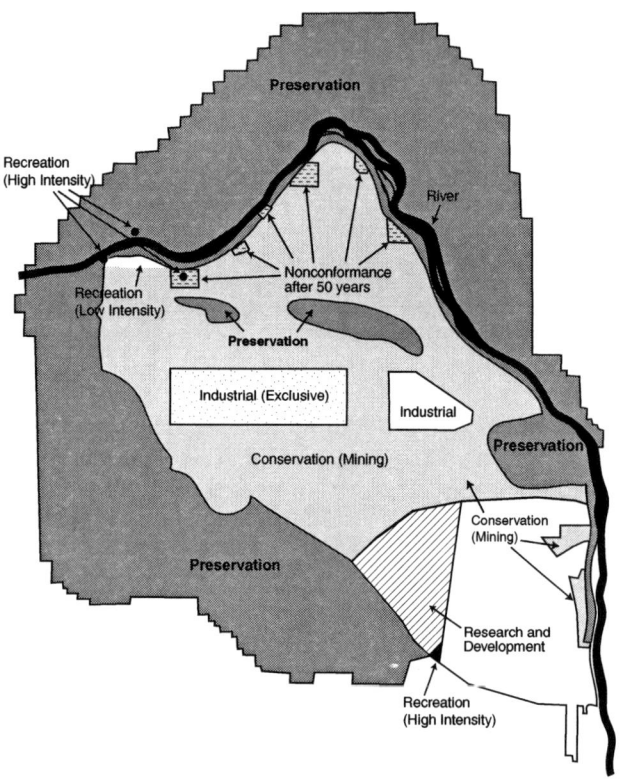

Figure 2.11 Future land use at the Hanford Site. SOURCE: DOE, 1999a, Figure 3.3.

Hanford Site Background 37

and cleanup activities planned in this area are approximately the following:[24]

- Spent nuclear fuel, special nuclear materials, and cesium and strontium capsules will be shipped off-site to a geologic repository for disposition.
- All retrievable transuranic solid waste will be shipped to the Waste Isolation Pilot Plant in New Mexico.
- High-level waste in the 200 Area tanks will be retrieved, immobilized in glass, and eventually shipped to a geologic repository. The low-activity radioactive waste streams created by processing this high-level waste will be immobilized and disposed of on site.
- Soil and groundwater contamination from past tank leaks and leaks during the waste retrieval process, as well as any waste remaining in the tanks after retrieval, the tanks themselves, and ancillary equipment (e.g., piping and diversion boxes) will remain in place. It is likely that surface caps and barriers will be placed over tank farms.
- Solid waste burial sites (including the ERDF) containing transuranic and low-level radioactive waste will remain in place and will be covered with surface caps and barriers.
- Vadose zone contamination from liquid discharges will mostly remain in place and be covered with surface caps and barriers.
- Facilities, with the exception of chemical processing facilities ("canyons"), will be torn down, and some may be covered with surface caps or barriers.
- Canyons with significant amounts of fixed contamination will be left in place and covered with a surface cap or barrier.

Currently, "there is no single collection of DOE documents that constitute (or identify fully) the approved post closure end state" for the Hanford Site (DOE, 1999g, p. 2.3).[25] To date, some end states for individual areas within the site have been established and are detailed in various environmental impact statements, environmental assessments,

---

[24]This information was provided in writing to the committee by the Integration Project after the committee's second meeting.
[25]The term *end state* is used to denote the condition of the site after DOE cleanup is completed. The end state can be characterized in terms of acceptable levels of residual contaminants or permissible site uses. The term is used in both its singular and its plural forms—for example, to refer to the overall end state for the Hanford Site or to the end states for specific regions or facilities within the site. See also *An End State Methodology for Identifying Technology Needs for Environmental Management, with an Example from the Hanford Site Tanks* (NRC, 1999a).

strategic plans, and records of decision (e.g., DOE, 1998b, 1998e, 1999a, 2000f; see also the Hanford Strategic Plan at http://www.hanford.gov/hsp/). However, several end states are not fully agreed upon, particularly in the 200 Area. For example, end states for groundwater remediation, high-level waste tank closure, and other facility closures (e.g., closure of the chemical processing facilities) have not yet been established. Also, final cleanup levels have not been determined for much of the waste to be permanently disposed of in the 200 Area.

The Columbia River comprehensive impact assessment (Kincaid et al., 2000, p. 3-3) coined the phrase "Hanford Site Disposition Baseline" (HSDB) to describe the suite of disposal and remedial actions that will occur as the Hanford Site moves towards closure. *Accelerating Cleanup: Paths to Closure* (DOE, 1998b, p. ES-3) states that "where decisions have not yet been made, sites make assumptions (e.g., site planning end states) about how those cleanup actions might be carried out so that sites can define work and develop schedule and cost estimates."

An initial statement of the Hanford Site Disposition Baseline (HSDB-2000) is available, and it consists of three tables covering the 100 Area; the 300, 400, and 600 Areas; and the 200 Area (Kincaid et al., 2000). The tables list the material type requiring remediation, the corresponding HSDB assumptions, and data needs. A similar set of tables is available for the same three areas, which are titled "Identification of Differences and Issues for Material Type and Areas at Hanford."[26] This has the advantage of referring to the Hanford Strategic Plan and to the environmental impact statements, environmental assessments, and records of decision to distinguish between disposition agreements, requirements, and assumptions. It also includes a summary of key differences among available documents and key issues.

DISCUSSION

The committee recognized early on in its information-gathering meetings that the absence of a clearly articulated end-state vision for the Hanford Site made it difficult to obtain a clear understanding of the exact nature and timing of future cleanup decisions. The lack of clearly defined decision points and options also makes it difficult for the Integration Project to develop an S&T program that is focused on filling well-defined knowledge gaps required to support well-defined site decisions, as detailed later in this report.

---

[26]These tables were provided to the committee by the Integration Project after its second meeting.

Hanford Site Background

Nevertheless, the Hanford Site Disposition Baseline and associated documents are, in the committee's view, very important to the S&T program because they indicate the general direction of work at the site and the kinds of knowledge gaps that may be important. In turn, this suggests the generic types of S&T that may be useful.

- These documents raise key issues for the S&T program—for example, How clean is clean enough? especially as applied to the need to retrieve 99 percent of the waste from the high-level waste tanks as currently stipulated in the Tri-Party Agreement. S&T could help formulate logical scientific and technical approaches for resolving these sorts of issues.
- They allow gaps in site remediation programs to be identified so that S&T efforts can be focused. For example, there is no mention of long-term stewardship (see Chapter 1) in the baseline, and in other documents, stewardship is restricted to 50 to 75 years. Stewardship in the context of these documents does not deal with long-term degradation of facilities and barriers, particularly in the 200 Area, which could require S&T to develop a robust monitoring and maintenance capability to ensure the long-term stability of the site.
- They indicate that the number of material dispositions not currently agreed upon is rather large. Agreements are being reached one at a time. A system that generically addresses the concerns of site stakeholders using logical, scientifically based information could help accelerate these decisions. If properly focused and timed, S&T could play a key role in resolving these issues by providing a technical basis for decision making by participating regulators and stakeholders.

These documents are also valuable because they provide useful guidance to the Hanford S&T programs in progress. They highlight key issues that require resolution (e.g., decisions concerning material dispositions and end states not yet agreed upon) and potential knowledge gaps to be addressed by S&T. Planning end points and planning end states will no doubt continue to evolve with time as S&T results become available and remediation progresses, which in turn will influence the future course of S&T. This interplay between the cleanup program and S&T is discussed in more detail in Chapter 10.

3
Overview of the Integration Project

The objective of this chapter is to provide an overview of the Integration Project to set the stage for the detailed assessments of the science and technology (S&T) program in subsequent chapters of this report. The committee relied on several key documents in preparing this chapter, most notably DOE (1998a, 1998d, 1999b, 2000a) and GAO (1998).

BACKGROUND AND HISTORY

The Groundwater/Vadose Zone Integration Project[1] was established in late 1997 in response to pressure from the U.S. Congress and Department of Energy (DOE) Headquarters for more effective coordination of the numerous waste management and clean-up efforts under way at the Hanford Site (DOE, 1998c). As discussed in Chapter 1, the integration effort grew out of investigations begun in 1994 to map radionuclide distributions around and beneath the single-shell tanks in the SX Tank Farm in the 200 Area at the site (see Chapter 2).[2] These investigations suggested that significant radionuclide migration into the deep vadose zone had occurred and that radionuclides had in fact reached groundwater in at least one instance. This discovery contradicted long-enunciated DOE assertions that radionuclides would not migrate to groundwater for thousands of years. Concurrent work by Los Alamos National Laboratory scientists suggested that leaks from the single-shell tanks in one tank farm may have been several times greater than previously reported (Agnew and Corbin, 1998, Table 2).

At the time the Integration Project was established, three organizations were responsible for waste management and cleanup at the Hanford Site. The work done by each of these offices was carried out by several private contractors with oversight by federal employees.

---

[1]The Groundwater/Vadose Zone Integration Project is referred to as the "Integration Project" in this and subsequent chapters.
[2]A good discussion of the events leading up to the formation of the Integration Project is provided in GAO (1998).

Overview of the Integration Project 41

1. The Tank Waste Remediation System Program was responsible for management and cleanup of the tank farms and underlying vadose zone.
2. The Office of Environmental Restoration was responsible for cleaning up the site, including the vadose zone and groundwater outside the tank farms.
3. The Office of Waste Management was responsible for managing stored and future-generated waste.

The Integration Project was overlaid onto these three existing organizations, and it was charged with coordinating the activities of these organizations with respect to investigations of the vadose zone, groundwater, and Columbia River.[3] The three organizations signed a memorandum of understanding in 1997 that outlined their responsibilities for the vadose zone at the site. The Environmental Restoration Program was directed to be the lead in this effort, and its contractor, Bechtel Hanford, was directed to take the lead in developing a plan of work. A draft of this plan was issued in December 1998 (DOE, 1998d), and updates of parts of the plan have been issued since that time (DOE, 1999b, 2000a).

The names of the three organizations responsible for waste management and cleanup at the site were changed in 1998 and 1999, but their responsibilities remain much the same:

1. The Office of River Protection, which was created by Congress in 1998, is now responsible for management and cleanup of the tank farms and underlying vadose zone.
2. The Office of Project Completion, Richland Office, is responsible for cleaning up the remainder of the site.
3. The Office of Integration and Disposition is responsible for managing stored and future-generated waste.

CH2M Hill is the primary contractor for the tank farm work, Bechtel Hanford is the primary contractor for the environmental restoration program, and Fluor Daniel Hanford is the primary contractor for nuclear materials management at the site. Table 3.2 provides a summary of the projects under these offices. Additional details are provided later in this chapter.

---

[3]The name "Groundwater/Vadose Zone Integration Project" does not reflect the potentially important role this project plays in protecting the Columbia River. The committee was told that the project name was coined in its early development stages, before its full scope was understood. By the time the full scope was established, the project name had become institutionalized.

It was clear even from an early draft of the Integration Project plan that the project scope was broader than suggested by its name. This is perhaps best illustrated by the mission statement in the December 1998 draft of the project specification (DOE, 1998d, p. 1-2):

> To ensure that Hanford Site decisions are defensible and possess an integrated perspective for the protection of water resources, the Columbia River environment, river-dependent life, and users of the Columbia River resources, the mission of the Groundwater/Vadose Zone Project is to develop and conduct defensible assessments of the Hanford Site's present and post-closure cumulative effects of radioactive and chemical materials that have accumulated throughout Hanford's history (and which continue to accumulate). To support this mission the Groundwater/Vadose Zone Project will also define those actions necessary to establish consistency and maintain mutual compatibility among site-wide characterization and analysis tasks that bear on decisions, receptor impact, and regulatory compliance. The Groundwater/Vadose Zone Integration Project will identify and oversee the science and technology initiatives pursued by the national laboratories (as necessary) to enable the assessment mission to be successfully completed.

As noted in Chapter 1, the main objectives of the Integration Project as outlined in this December 1998 draft are as follows:

1. Integrate all Hanford Site groundwater/vadose zone related work scope.
2. Predict current and future impacts resulting from contaminants that have been (or are predicted to be) released to the soil column at the Hanford Site.
3. Provide a sound science and technology basis for site decisions and actions.
4. Promote open and honest involvement by Tribal Nations, regulators, and other stakeholders so that project outcomes reflect expressed interests and values.
5. Establish an independent technical peer review.

The Integration Project has both technical and nontechnical objectives. On the technical side, the Integration Project is responsible for promoting the use of sound science and technology in decision making at the site. The project is also responsible for promoting interactions with outside parties who have an interest in Hanford so that local interests and values are taken into account in those decisions.

The third and fifth objectives are particularly germane to this National Research Council study. As noted in Chapter 1, the study was requested by DOE Headquarters as part of the site's efforts to obtain independent technical reviews of its programs. Also as noted in Chapter 1, the committee has been asked to review the S&T work that is occurring under the Integration Project and to offer recommendations to improve its technical merit and applicability to site cleanup decisions. A brief review of the science and technology element of the Integration Project is provided below. More details are provided in subsequent chapters.

SCIENCE AND TECHNOLOGY PROGRAM

The objective of the Integration Project's science and technology program is to provide the data, tools, and understanding to predict present and future impacts and to promote sound decision making (Sidebar 3.1). The Integration Project's science and technology program is organized into the six technical elements listed below. Within each of these technical elements, the Integration Project supports (or plans to support) scientific and technical studies to improve the understanding of contaminant inventories, locations, fate and transport processes, and impacts on the Columbia River.

1. The *Inventory Technical Element* supports studies to develop improved estimates of chemical and radionuclide inventories at the Hanford Site, especially for wastes disposed of or discharged to the subsurface. There are a number of site databases that track waste inventories, most notably the Hanford Environmental Information System, Waste Inventory Data System, and Solid Waste Inventory Tracking System (see Chapter 5). However, the data in these systems are incomplete, primarily because waste inventories were not tracked very carefully during much of the site history (see Chapter 2 for a more detailed discussion).

2. The *Vadose Zone Technical Element* supports studies to develop a better understanding of subsurface contaminant behavior in the vadose zone—for example, studies to improve the understanding of fate and transport processes in the vadose zone, studies to improve conceptual and numerical models of contaminant fate and transport in the vadose zone, and studies to test advanced characterization tools and methods.

3. The *Groundwater Technical Element* supports studies to improve site-wide assessments of contaminant fate and transport in groundwater at the site—for example, studies to improve modeling of contaminant fate and transport in groundwater and studies to improve the understanding of contaminant locations in the subsurface and of locations and fluxes of contaminant releases to the Columbia River.

4. The *Columbia River Technical Element* supports studies to provide an enhanced understanding of the potential impacts and consequences of contaminant releases to the Columbia River—for example, studies to improve conceptual models of the river, studies to parameterize fate and transport models, and studies to improve the understanding of the effects of contaminants on riverine biota.

5. The *Risk Technical Element*, which is still under development, will focus on improving the understanding of risks, broadly construed,[4] posed by contaminant migration at the site and on reducing uncertainties in risk assessment methodologies.

6. The *Remediation and Monitoring Technical Elements*, which have not yet been developed, will focus on improving capabilities to remediate and monitor environmental contamination at the Hanford Site.

SCIENCE AND TECHNOLOGY PROGRAM PLANNING THROUGH RESEARCH AND DEVELOPMENT "ROADMAPS"

Problems to be addressed by the six technical elements listed above are being identified through a process that DOE calls *research and development (R&D) roadmapping*.[5] In DOE parlance, a *roadmap* is an R&D plan developed to address explicitly posed technical problems and to guide investment decisions so that the needed R&D work can be completed in time to make critical site decisions. The roadmap itself is a document that identifies the technical problems to be addressed by R&D, with a plan that lays out objectives, priorities, schedules, and budgets for addressing them. A roadmap is usually developed through a series of meetings or workshops that bring together experts who understand the problems that must be addressed (problem holders), experts who understand how to address these problems (problem solvers), and other parties who have an interest in the work to be done (stakeholders).

---

[4]The Risk Technical Element considers ecological, human health, economic, and sociocultural impacts, the latter two of which are not usually considered in standard risk assessments.

[5]The roadmapping concept originated in industrial R&D labs and has been embraced by DOE for many of its R&D programs through the strong encouragement of Ernest Moniz, who served as DOE Under Secretary at the time the Integration Project was created.

Overview of the Integration Project 45

SIDEBAR 3.1 What is Integration Project S&T?

During the course of this study, the committee neither found in the written documentation it reviewed nor heard in the oral briefings it received from Integration Project staff a definition of *science and technology* in the context of the Integration Project Roadmap (DOE, 1999b). The roadmap describes the objectives of S&T—"to provide new knowledge, data, tools, and the understanding needed to enable the Integration Project's mission"—and also notes that S&T "is focused on resolving key technical issues that help inform and influence decisions," but it does not describe what the Integration Project considers to be within the scope of S&T. Moreover, the core projects also fund and carry out a significant portion of the S&T effort at the site, and their definitions may not be consistent with those used by the Integration Project.

Consequently, the committee has applied what it considers to be the generally accepted definitions of science and technology in reviewing the S&T program: *science* is the discovery of knowledge, especially as obtained and tested through scientific methods, whereas *technology* is the application of scientific knowledge to particular problems. The Integration Project's focus on both science *and* technology provides unique and important opportunities to weave together knowledge creation, knowledge integration, and knowledge application to solve an important societal problem.

In fiscal year 1998, DOE held a series of workshops involving site contractors (problem holders), national laboratory scientists (problem solvers), and representatives of regulatory agencies, Tribal Nations, and other interested parties (stakeholders) to develop initial (Rev. 0) roadmaps for four of the six technical elements: inventory, vadose zone, groundwater, and Columbia River. These roadmaps are provided in *Groundwater/Vadose Zone Integration Project Science and Technology Summary Description* (DOE, 1999b). This document will be referred to as the Integration Project Roadmap in the remainder of this report.

In fiscal year 1999, DOE held additional meetings with staff from the DOE Center for Risk Excellence, national laboratory and university scientists, Tribal Nations, and other stakeholders to develop a roadmap for the Risk Technical Element. This roadmap and updated roadmaps for the other four technical elements are provided in "Rev. 1" of the roadmap document (DOE, 2000a). During the current (2001) and next (2002) fiscal years, DOE plans to develop additional roadmaps for the remediation and monitoring technical elements, presumably using the same process that was used to develop the other five roadmaps.

The Integration Project Roadmap (DOE, 2000a) describes R&D needs, products, schedules, and budgets. The roadmap descriptions are general in nature and provide little or no technical detail on individual S&T projects. This is a key document for the committee's review, and additional details of the roadmap are provided in subsequent chapters.

IMPLEMENTATION OF INTEGRATION PROJECT ROADMAP

The projects outlined in the Integration Project Roadmap are designed to provide scientific and technical information to meet DOE's cleanup or waste management objectives. To help ensure the timely delivery of useful information, the Integration Project has developed science-user teams for each of the technical elements discussed above. These teams comprise Integration Project staff, contractor staff from DOE's "core" remediation and waste management projects,[6] and national laboratory researchers. Some of the teams also involve principal investigators from Environmental Management Science Program (EMSP)[7] projects relevant to Hanford Site cleanup (these projects are discussed in more detail elsewhere in this report). The science-user teams are responsible for planning and implementing the R&D work and ensuring that the results are transferred to problem holders in a timely fashion.

The Integration Project Roadmap identifies projects that provide R&D support to five Hanford Site core projects as well as two Integration Project efforts:

- The *Tank Farm Vadose Zone Project* (core project) is responsible for remediating or stabilizing contaminants in the vadose zone beneath the 200 Area tank farms. Planning for this work is under way, but actual remediation has not yet begun.

---

[6]The core projects are responsible for the actual work done at the Hanford Site to remediate and/or stabilize waste and contaminants.
[7]The EMSP is a mission-directed, basic research program that provides three-year grants to researchers in national laboratories, academia, and industry. The grants are awarded based on competitive peer review that considers both scientific merit and relevance to DOE's cleanup needs. The program was established by Congress in 1996 and is managed jointly by DOE's Office of Science and Office of Environmental Management. See National Research Council (1997, 2000a) for a description of this program.

- The *Groundwater Project* (core project) is responsible for site-wide groundwater monitoring and remediation.
- The *200 Area Remedial Action Project* (core project) is focused on the remediation and/or stabilization of waste burial grounds and discharge sites in the 200 Area.
- The *River Monitoring Project* (core project) is responsible for monitoring the Columbia River to meet regulations and compliance agreements.
- The *Immobilized Low-Activity Waste Project* (core project) is responsible for development of a disposal facility for low-activity waste that will be generated during retrieval, processing, and immobilization of high-level waste from the 200 Area tank farms.
- The *System Assessment Capability Project* (SAC; Integration Project) is responsible for the development of models and databases that can be used to conduct site-wide risk assessments.
- The *Characterization of Systems Project* (Integration Project) is responsible for the development of data and conceptual models for the vadose zone, groundwater, and river.

The Integration Project's R&D activities take several forms. As shown in later chapters, most of the Integration Project's current R&D work is being conducted through the EMSP, a basic research program funded through DOE Headquarters. The Integration Project also provides direct funding for shorter-term, applied R&D work. Some additional R&D may be funded directly by the national laboratories through laboratory-directed research and development funds.[8] R&D work, whether under the auspices of the Environmental Management Science Program or the Integration Project, may be conducted in conjunction with core project activities. The Integration Project refers to R&D done in conjunction with core projects as "wrap-around science."

SCHEDULE AND BUDGET

Rev. 1 of the Integration Project Roadmap (DOE, 2000a, Figure 4.1 therein) provides a logic diagram of R&D activities that extends through fiscal year 2005, with notational lines to indicate that some work will extend beyond that date. The budget for the program (DOE, 2000a,

---

[8]Multiprogram DOE national laboratories are authorized by Congress to spend a percentage of their operating budgets on research and development activities "of a creative and innovative nature ... selected by the director of a laboratory for the purpose of maintaining the vitality of the laboratory in defense-related scientific disciplines" (National Defense Authorization Act for Fiscal Year 1991).

Table 5-1; see Table 3.1 in this report) extends through fiscal year 2004 (FY04) and indicates that the Integration Project's S&T effort will involve an investment of between about $1 million and $16 million per year to complete the planned work. This budget has been reduced since the roadmap was published, as noted by the bottom row of the table for

TABLE 3.1 Budget for the Integration Project's Science and Technology Program

Budget Element	Planned Fiscal Year Funding Levels[a] (thousand dollars)						Total
	1999	2000	2001	2002	2003	2004	
Inventory	130	410	845	130	130		1,645
Vadose zone	120	3,170	5,500	6,500	6,500	3,000	19,840
Groundwater			450	900	400	600	2,350
River		250	1,000	1,250	750	850	4,100
Risk			3,750	5,300	3,800		12,850
Remediation							
Monitoring							
Roadmap planning and implementation	900	900	500	500	500		3,300
Planned funding level[b]	1,150	4,730	12,045	14,580	12,080	4,450	51,985
Actual funding level[c]	1,333	4,700	4,600	—	—	—	—
Other S&T program funding levels[d]	24,000[e]			—	—	—	—

[a]The figures in this table represent Integration Project funding levels for the S&T program (Table 5.1 of DOE, 2000a). Additional funding for activities identified in the S&T roadmap is provided by other Hanford core projects as well as the Integration Project through its SAC and Characterization of Systems projects.
[b]The figures shown are calculated by summing the funding levels for each fiscal year.
[c]The actual funding levels are from DOE (2000c).
[d]The figures shown in this column represent planned funding for the Integration Project S&T program from other DOE sources, for example, the EMSP.
[e]The EMSP awarded funding to 31 projects. This funding will be distributed from fiscal year 1999 through fiscal year 2002.

FY01. The budget reduction is being achieved primarily by delaying planned work.

More detailed budgets for each of the technical elements shown in Table 3.1 are provided in the Integration Project Roadmap (DOE, 2000a) and are reproduced in Chapters 5-9 of this report. There are several inconsistencies between Table 3.1 and the budgets shown in the later chapters due to funding reductions and changes in budget priorities since the roadmap budgets were published. Nevertheless, the committee considers the Integration Project Roadmap budgets given in Chapters 5-9 to be important because they provide an indication of projected funding needs during the first five years of the project's existence.

DISCUSSION

This chapter provides an overview of the Integration Project to set the stage for the detailed assessments of the science and technology plan in subsequent chapters. The material in this chapter reflects the committee's understanding of the Integration Project's S&T program as it existed when the committee completed its information gathering in late March 2001.

Several preliminary observations are worth noting at this point. The Integration Project has been superimposed onto a number of preexisting, highly complex, multicontractor "core" waste management and cleanup projects at the site (see Table 3.2). The Integration Project has been given the challenging task of providing scientific and technical information to these preexisting projects, but it has very restricted authority and budget to carry out this mandate. It has direct control over only the small amount of money it distributes to the R&D effort each year (Table 3.1), and it has no authority over the clean up decisions to be made. It is not even clear in many cases who "owns" the Integration Project's R&D results. To add to this challenge, the core project missions themselves also appear to be changing as the end-state decisions to be made at the site (Chapter 2) are developed.

The Integration Project is operating in an unstable programmatic environment, which makes it difficult to plan an R&D program that meets site needs and schedules. Nevertheless, with cleanup work at the site planned to extend until at least 2046 (see Chapter 2), there would certainly appear to be ample opportunity to maintain an R&D effort that, through proper planning and focus, will fill critical knowledge gaps for the cleanup program at Hanford. Suggestions for how the Integration Project can operate more successfully in this unstable environment are given in Chapter 10.

TABLE 3.2 Core and Integration Project Responsibilities for Environmental Management at the Hanford Site

DOE Office		Project	Responsibility
Office of Project Completion, Richland Office	Integration Projects	GW/VZ Integration Roadmapping Project and S&T Elements	Plan and integrate S&T for environmental decision making; coordinate stakeholder involvement
		SAC Project	Models and databases for site-wide risk assessments
		Characterization of Systems Project	Models and data for vadose zone, groundwater, and Columbia River
	Core Projects	200 Area Remedial Action Project	200 Area disposal sites outside tank farms
		Groundwater Project	Site-wide groundwater monitoring and remediation
		River Monitoring Project	Monitoring the Columbia River
Office of Project Completion, River Protection Office		Tank Farm Vadose Zone Project	Unsaturated zone around tanks
		Immobilized Low-Activity Waste Project	Disposal of low-activity waste generated from tank waste immobilization operations

4
System Assessment Capability

As discussed in Chapter 3, one of the primary functions of the Groundwater/Vadose Zone Integration Project is to predict current and future impacts on humans and the environment resulting from the release of contaminants at the Hanford Site. The Integration Project is developing what it calls the *System Assessment Capability*, or SAC, to estimate these current and future impacts. The SAC will comprise a set of models and parameter databases that can be used to obtain quantitative estimates of cumulative impacts of contaminant releases on water resources, biological (including human) systems, cultures, and economies in the region around the Hanford Site extending over hundreds of years.

Although the SAC is not formally part of the Integration Project's science and technology (S&T) program, it is a potentially important end user of S&T products. These products include mass balances of inventories and contaminant releases[1] to be provided by the Inventory Technical Element (Chapter 5); conceptual models, numerical models, and parameter databases for contaminant fate and transport to be provided by the Vadose Zone, Groundwater, and River Technical Elements (Chapters 6-8); and human, ecological, economic, and cultural impact data to be provided by the Risk Technical Element (Chapter 9).

Given the importance of the S&T program to the SAC, the committee provides a short review and assessment of the SAC in this chapter to set the stage for more detailed assessments of the S&T technical elements later in this report. The primary purpose of this assessment is to identify knowledge gaps that, if addressed through additional S&T work, could improve the usefulness of the SAC as a predictive tool.

---

[1]The term *inventory* is used by the Integration Project to describe the quantities of radionuclides and chemicals that have been placed in storage and disposal facilities at the Hanford Site—for example, high-level radioactive waste placed in underground tanks or transuranic waste disposed in near-surface trenches. As noted in Chapter 2, some of this waste has migrated out of these disposal facilities and into the vadose zone or groundwater. The committee uses the term *contaminant release* to describe these releases, whether accidental or intentional.

SCOPE OF THE SYSTEM ASSESSMENT CAPABILITY

The SAC is designed to predict contaminant migration through the vadose zone, groundwater, and Columbia River and its impacts on a variety of receptors, using as a starting point historical waste inventories from Hanford operations. A conceptual illustration of the SAC is provided in Figure 4.1. After historical inventory inputs are prescribed, SAC uses three sets of numerical codes to estimate contaminant migration and impacts: the first to simulate the release of radionuclide and chemical inventories into the environment; the second to simulate contaminant migration through the environment; and the third to estimate risk and impacts from this contaminant migration. Figure 4.1 is a very simplified conception of the SAC—the actual model consists of more than a dozen modules and data interfaces (Figure 4.2) designed to run on a network of computer workstations.

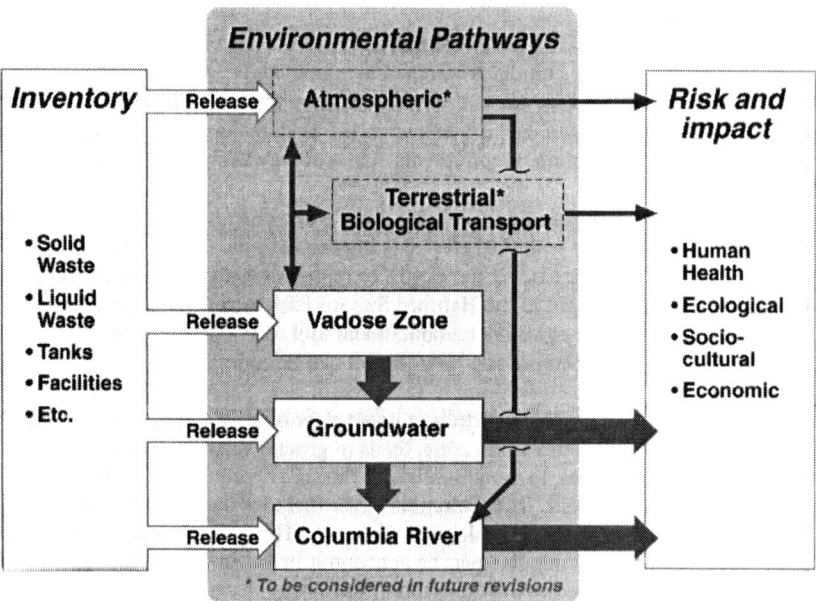

Figure 4.1 Conceptual illustration of the System Assessment Capability. SOURCE: Kincaid et al., 2000, Figure 1-1.

System Assessment Capability

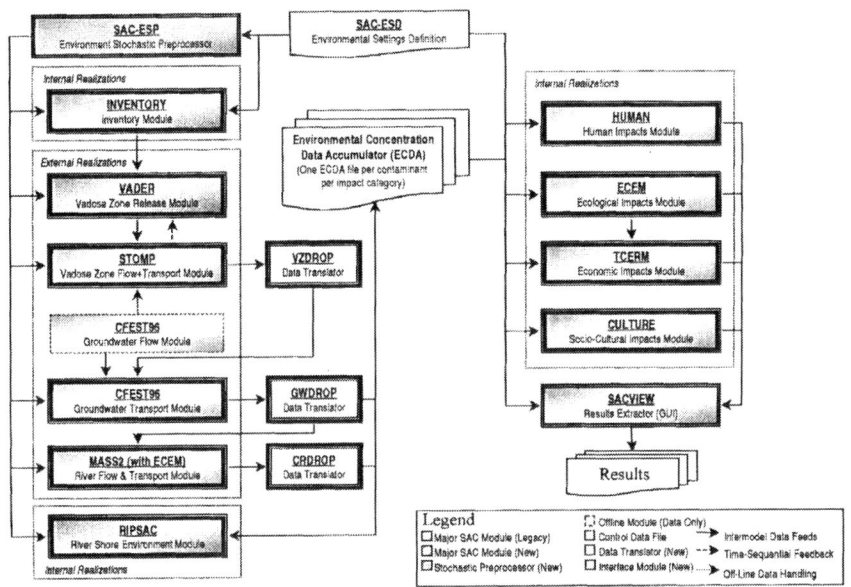

Figure 4.2 Numerical code modules and data interfaces for SAC Rev. 0. SOURCE: Kincaid et al., 2000, Figure 5-1.

The SAC is complex in both design and implementation. It will attempt to predict cumulative impacts to multiple receptors at multiple locations from multiple source terms[2] for 1,000 years. A broad range of impacts will be estimated, including ecological, economic, and sociocultural effects.

The Integration Project will produce several revisions of the SAC, each having progressively greater capabilities and complexity, during the next several years. SAC Rev. 0 is intended to be a proof-of-principle implementation. It consists of a set of sophisticated codes run in a

---

[2]A *source term* comprises the locations and quantities of radionuclides and chemicals that are available to be transported to the environment from facilities and waste sites.

simplified manner to demonstrate that an assessment can be conducted at the site-scale level. SAC Rev. 1 is intended to be a production implementation that is sufficiently developed to support site decisions. It is planned to be used, for example, to conduct Comprehensive Environmental Response, Compensation, and Liability Act of 1980 (CERCLA)[3] reviews, analyze impacts of changes to cleanup baselines, and develop closure plans for the 200 Area tank farms (see Chapter 2). At least two additional versions of SAC (i.e., Rev. 2 and 3) are planned in the Integration Project baseline through 2005 (DOE, 2000a). These versions will continue to add capabilities, complexity, and refinements to Rev. 1.

The Integration Project was in the process of completing SAC Rev. 0 during the committee's information-gathering meetings, and the committee received documents (e.g., BHI, 1999; Kincaid et al., 2000) and briefings (Appendix B) on the details of this version. A summary of the main design elements of Rev. 0 is provided in Table 4.1. As shown in Table 4.1, SAC Rev. 0 is designed around a set of simplifying and limiting assumptions about contaminant transport and long-term site conditions:

- Contaminants will be released from a limited number of "representative" locations at the site. The Waste Information Database System (see Chapter 5) lists more than 2,600 release locations across the site.[4]

- Transport through the vadose zone is modeled as a one-dimensional, homogeneous, isotropic continuum, and transport through the saturated zone as a two-dimensional phenomenon. Actual contaminant transport in the vadose zone and saturated zone occurs in three dimensions, and potentially important subsurface transport features (such as clastic dikes) cannot be modeled in the Rev. 0 configuration.

- Site conditions, including climate, river discharges, and economic and sociocultural conditions, will be unchanged from the present through the year 3050. Rev. 0 also ignores the potential impacts of extreme events such as large fires, floods, and seismic events.

---

[3]Most of the cleanup work at the Hanford Site is proceeding under CERCLA, which requires periodic reviews to ensure that cleanup objectives continue to be met.

[4]Initially, the Integration Project planned to use eight different representations of the vadose zone to model these releases: one each in the 100 and 300 Areas and six representations for the Central Plateau in and around the 200 Area. The number was later increased to 13 to add more representations for the 100 Area. The 2,600+ waste sites at the Hanford Site are to be aggregated into 50-100 representative release sites in the Rev. 0 assessment. This number will probably increase to 500 before this assessment is completed (Mark Freshley, Pacific Northwest National Laboratory, written communication, April 12, 2001).

TABLE 4.1 Principal Design Elements of SAC Rev. 0

Element	Description	Treatment of Uncertainty[a]	Comments
Contaminants modeled	Eight radionuclides and three chemicals: tritium, technetium-99, iodine-129, strontium-90, Cs-137, plutonium-239/240, uranium (as both a chemical and radionuclide), carbon tetrachloride, and chromium		SAC developers consider these the most significant contaminants at the site and potentially significant future dose-risk contributors (see DOE, 1999c)
Contaminant pathways modeled	Facility release-vadose zone-groundwater-Columbia River		Atmospheric and terrestrial transport pathways are ignored
Assumed site end states	Hanford Site Disposition Baseline (see Chapter 2)		Alternates to this baseline will not be examined
Contaminant release sites modeled	Eight release sites: one in the 100 Area; one in the 300 Area; six on the Central Plateau in and around the 200 Area		See footnote 4 in this chapter
Model temporal domain	1944-3050		SAC will be run for a 1,000-year interval after assumed site closure in 2050

Table 4.1 Continued

Element	Description	Treatment of Uncertainty[a]	Comments
Model spatial domain	Columbia River from Priest Rapids Dam to McNary Dam and the Hanford Site from Rattlesnake Mountain to the Columbia River		The upstream model boundary of the Columbia River has been moved to the Vernita Bridge, which is located just upstream of the Hanford Site boundary, to save computational time
Inventory model	Derived from Hanford Defined Waste Model (Agnew, 1997), other inventory databases, and identification of waste sites based on waste stream and process knowledge. Tank leak estimates will be extrapolated to individual facilities (release points) to estimate leak volumes and masses as a function of time	Will be estimated using expert elicitation to evaluate the quality of the recorded data coupled with Monte Carlo estimates constrained by the mass balance for total inventory. Results will be compared qualitatively with the timing and magnitude of releases observed over the past 55 years of field operations	It is not clear how qualitative comparisons will be performed or how uncertainty will be characterized from such comparisons. Inventory uncertainty is inherently higher for individual contaminant sources than for groups of sources.

Contaminant release models	Both "pass-through" and waste form dissolution models are used	Output distributions for the release models will be estimated using Monte Carlo simulations with predefined input distributions for model parameters. Results are to be compared qualitatively with the timing and magnitude of releases (e.g., breakthroughs of contaminants to the water table) observed during the past 55 years of field operations	The committee was unable to find an explanation of how these input parameter distributions will be derived. It is not clear how qualitative comparisons will be performed or how uncertainty will be characterized from such comparisons
Vadose zone model	Vertical transport through a four- to six-layer homogeneous and isotropic continuum using a linear sorption isotherm model to simulate geochemical reactions	Output distributions for the vadose zone models will be defined using Monte Carlo simulations with predefined input distributions and model parameters. Results will be compared qualitatively with the timing and magnitude of releases (e.g., breakthroughs of contaminants to groundwater) observed during the past 55 years of field operations	Model incorporates radioactive decay for radionuclides and 'pseudodecay' for chemicals. It is not clear how qualitative comparisons will be performed or how uncertainty will be characterized from such comparisons

Table 4.1 Continued

Element	Description	Treatment of Uncertainty[a]	Comments
Groundwater model	Flow is modeled with the Hanford Site unconfined aquifer model (Cole et al., 1997, in Kincaid et al., 2000) run in a two dimensional mode, using a linear sorption isotherm model to simulate geochemical reactions	Only a single conceptual model will be evaluated, and a limited analysis of parameter uncertainty will be considered in Rev. 0 of the model	Model incorporates radioactive decay for radionuclides and 'pseudodecay' for chemicals. Details of the analysis of parameter uncertainty are not provided
Columbia River shore model	Contaminant concentrations within the riparian zone are derived from groundwater and river concentrations at the aquifer-river boundary using results from the groundwater flow model and the Columbia River model	Represented by the uncertainty in empirical coefficients defining the relationships between contaminant concentrations in groundwater, bank seepage, and sediments	Model incorporates flow reversals between the river channels and groundwater due to fluctuating river stages

Columbia River model	Two-dimensional (bank-to-bank), depth-averaged flows of water, sediment, and contaminants are modeled, as are biological transport and food web transfers. Solid-aqueous distribution coefficients and empirical dilution factors are used to obtain estimates of radionuclide and chemical concentrations in river water	Will be represented by the uncertainty of the empirical parameters for each submodel and input to the river environment. A detailed analysis of uncertainty will be deferred to Rev. 1	Emphasis is on the river-groundwater interface. Details of the analysis of parameter uncertainty are not provided
Spatial resolution (size of computational grid)	375 meters by 375 meters for the groundwater model		SAC stores contaminant concentrations at 2,500 locations in groundwater for development of spatial distribution maps and ecological assessments at 200 locations in the environment for the development of temporal profiles

Table 4.1 Continued

Element	Description	Treatment of Uncertainty[a]	Comments
Temporal resolution	1- to 5-year intervals for the period up to closure and whatever intervals are needed to maintain control accuracy thereafter		
Model results	Results based on 100 realizations		Model results may be based on fewer than 10 realizations to save computing time
Global assumptions	Climate remains unchanged through 3050		Model uses 1961-1990 climate data as a baseline
	Upstream reservoir systems are maintained through 3050		Model uses 1944-present Columbia River discharge data as a baseline
	Human populations, economic conditions, and cultures remain unchanged through 3050		Model assumes that conservation and preservation (Chapter 2) will continue to be the dominant site land uses
	Extreme events (e.g., floods, fires, earthquakes) are not considered		

| Impact metrics produced by the model | Ecological risk, human health risk, economic impacts, and sociocultural impacts |

[a]In many cases, there was no documentation on how uncertainty is to be treated in various elements of the SAC. In these cases, no entry is made in this column of the table.

SOURCE: DOE, 2000a, supplemented by Integration Project reviews of this table.

- Impacts are estimated only for the first 1,000 years following site closure. It is not clear that the time to peak risk will occur in the first 1,000 years, however, especially in light of the long half-lives of some key radionuclides and expected long travel times to receptors of interest. For example, technetium-99, iodine-129, and uranium-238 have half-lives of 0.2 million, 16 million, and 4.5 billion years, respectively. If peak risk occurs beyond 1,000 years, then other model assumptions, particularly the assumption that climate remains unchanged, may not be realistic.

SCHEDULE AND BUDGET

Work began on SAC Rev. 0 in fiscal year 1999 and is planned to be completed by the end of fiscal year 2001, with the initial model runs scheduled to be completed by the end of July 2001. Model run times are very long, however, and the Integration Project has had to cut the planned number of realizations to 10 to keep the project close to schedule (see comments in Table 4.1). It is not clear what will be learned from the small number of planned realizations, except to demonstrate that the model can produce a numerical "answer." The small number of realizations seems inadequate to capture the behavior of the system.

Work on SAC Rev. 1 was initially scheduled to begin around the end of fiscal year 2001. However, in the latest update issued by the Integration Project, the schedule for Rev. 1 has slipped to fiscal year 2002 (DOE, 2000c, p. 20). According to the Integration Project Roadmap, subsequent revisions to SAC are scheduled to be produced at 18-month intervals (DOE, 2000a, Figure 4-1) culminating in the release of Rev. 3 in 2005.

The Integration Project is now considering an alternative schedule that would involve enhancing Rev. 0 and using it to perform several alternative assessments during fiscal year 2002. The Integration Project would then solicit feedback from the Department of Energy (DOE), regulators, and stakeholders about how the capability should evolve and, if appropriate, will produce a Rev. 1 before 2005.

Funding for the SAC is provided through the DOE-Richland Office budget. SAC received $1.9 million in fiscal year 1999 and $2.85 million in fiscal year 2000. The fiscal year 2001 budget request was $2.0 million, but only $1.7 million was allocated.[5]

---

[5]In response to a $300,000 cut in the SAC budget for fiscal year 2001, the Integration Project plans to delay completion of the documentation of Rev. 0 results until fiscal year 2002.

System Assessment Capability

DISCUSSION

As a proof-of-principle implementation, SAC Rev. 0 is not intended to make specific predictions of contaminant transport and its impacts at the Hanford Site. That capability is planned for subsequent revisions. Moreover, it is not clear how "proof of principle" will be demonstrated, given that both the model results (see last column of Table 4.1) and the historical release data to which it will be compared have high degrees of uncertainty.

For SAC to achieve a reliable, predictive capability, however, additional S&T will be needed to close several important knowledge "gaps." The gaps that the committee judges to be most important are described briefly in the following paragraphs. More details are provided in subsequent chapters.

1. A lack of data on the three-dimensional distributions of contaminants in the vadose and saturated zones at the Hanford Site will greatly limit the ability to calibrate or validate the SAC as a reliable risk assessment tool. Relatively few data sets are available on contaminant distributions, concentrations, and speciation in the unsaturated zone deeper than 20 to 30 meters (Mark Freshley, Pacific Northwest National Laboratory, written communication, September 6-8, 2000; Myers and Gee, 2000). In addition, the lateral extent of contaminants beyond the "footprints" of waste disposal sites, such as tanks, cribs, and trenches, is poorly known. Absent knowledge of the current state of contamination in the subsurface, it will be difficult if not impossible to assess the reliability of contaminant migration and impact predictions derived from the SAC.

Equally important, it will be necessary to characterize the uncertainty associated with both the model predictions and the measured distributions of contaminants in the subsurface to which those predictions are to be compared. To this end, new procedures will have to be developed or adapted to characterize uncertainties and perform these comparisons in a manner that allows one to determine the degree of "success."

2. A lack of understanding of the three-dimensional nature of contaminant transport will limit the ability of SAC to provide accurate estimates of residence times for contaminants in the vadose and saturated zones. As discussed in Chapter 6, field experiments at Hanford have demonstrated clearly that fluid transport in the vadose zone is fully three dimensional. The three-dimensional nature of contaminant transport in the vadose zone is also illustrated in the document *Preliminary System Assessment Capability Concepts for*

Architecture, Platform, and Data Management (BHI, 1999). Similarly, groundwater measurements have shown that large vertical gradients can exist in contaminant concentrations in the saturated zone.

A separate but related problem is that hydraulic and transport parameters used in the transport models are derived from laboratory measurements on centimeter-scale core samples and are extrapolated to scales relevant to field transport. The scientific basis of an "upscaling" algorithm to calculate "effective" parameters for a large block of heterogeneous sediments from highly variable measurements on small samples has not been demonstrated. This problem is discussed in more detail in Chapter 6 and Appendix C.

3. A lack of understanding of the effects of extreme (high-magnitude and low-frequency) events such as large fires, floods, and earthquakes will limit the ability of SAC to provide accurate estimates of contaminant movement over time scales during which wastes will remain hazardous. Although the probability of occurrence of such extreme events in a single year is low, the consequences of these events could be much higher than those predicted solely by the advective-dispersive transport mechanisms considered in the SAC, especially over the time scales during which wastes will remain hazardous—typically on the order of 10^3 to 10^5 years.

The Hanford Site is a fire-prone ecosystem,[6] as evidenced by range fires in 1984 and 2000, each of which burned about half of the area of the site.[7] Fire represents a potentially important agent for mobilizing contaminants contained in vegetation and near-surface soils. Radionuclides contained in the burned vegetation can be released directly into the atmosphere, and near-surface contaminants could be mobilized by increased infiltration or surface erosion accompanying the loss of vegetation. The effects of fire on vegetation removal may be magnified during periods of severe drought, when a lack of precipitation would inhibit the recovery of burned areas.

The Hanford Site is also vulnerable to different types of flooding events, ranging from failures of pressurized water mains to catastrophic flooding. The latter has occurred repeatedly during glacial periods in the last 100,000 years. Even under current (interglacial) climatic conditions, intense rainfall occasionally saturates the land surface and generates intense runoff events with attendant sediment transport. Such flooding

---

[6] In fact, the ecosystem is structured by fire.

[7] It can be argued that such range fires have return periods on the order of decades and are not extreme events when measured against time scales of waste hazards at the site. Nevertheless, this does not diminish their potential importance as a contaminant transport agent.

System Assessment Capability 65

events could potentially result in the erosion and transport contaminants from near-surface soils and waste burial sites.

These extreme processes will have to be better understood and incorporated into later revisions of the SAC if it is to provide reliable long-term estimates of contaminant transport at the site. Understanding the potential impacts of such events is essential for making informed and durable site cleanup decisions and siting new waste disposal facilities. This issue will be discussed in more detail in Chapter 9—see especially Sidebar 9.1.

4. Exposure pathways other than through groundwater may exist at the Hanford Site—these are not considered in the current version of the SAC. For example, exposures to surface contamination may occur as a result of burrowing animals or erosion, or such contamination may exist in previously undetected locations. Humans could be exposed through dust inhalation, soil contact, or consumption of contaminated animals. Depending on what assumptions are made about groundwater use at Hanford, the soil contamination pathway could be significant from a risk perspective. Although additional exposure pathways may be included in later versions of the SAC, the Integration Project S&T program does not appear to be designed to support such additions. This issue is discussed in Chapter 9.

A common theme that cuts across these knowledge gaps is *uncertainty*. Characterizing uncertainty in the SAC, both in general terms and for specific applications, will be a difficult but essential task for the Integration Project. Equally important will be the adaptation of appropriate statistical tools that allow reasonable conclusions to be drawn even in light of such uncertainties. The committee believes that S&T can play a central role in reducing uncertainty—for example, through the collection of data on current contaminant conditions at the site and the development or adaptation of procedures to validate SAC predictions. This S&T work must be conducted concurrently with SAC development, so that results from current versions of SAC can be interpreted properly and S&T results can be incorporated into future revisions.

5
Inventory Technical Element

As noted in Chapter 3, the Integration Project's Inventory Technical Element supports studies to develop estimates of chemical and radionuclide inventories[1] at the Hanford Site. The radionuclide inventory includes any radioactive material imported to or produced at Hanford with a half-life greater than 5 years and activity in excess of 1 curie. The chemical inventory includes chemicals imported, manufactured, or produced at Hanford and other chemicals identified in the monitoring or characterization programs. The "other" category includes new chemicals produced, for example, through biological degradation of existing chemicals in the environment.

There are more than a dozen databases maintained by the Hanford Site and a large number of Hanford Site documents that contain chemical and radionuclide inventory information. The primary inventory-related databases include the following:

- The *Waste Inventory Data System* (WIDS) contains information on more than 2,500 potential waste sites at Hanford. The database tracks descriptions of the sites, their locations, and sampling or testing information.
- The *Solid Waste Inventory Tracking System* (SWITS) tracks inventories on radioactive and nonradioactive solid waste generated on-site and imported from off-site facilities.
- The *Hanford Environmental Information System* (HEIS) contains Hanford Site environmental sample data, including data from groundwater, waste sites, and soils.
- The *Tank Characterization Database* (TCD) contains tank waste analytical data, historical data, and surveillance data.
- The *Track Radioactive Component* (TRAC) database contains modeled estimates of tank waste radionuclide inventories. A more recently developed database, the *Hanford Defined Wastes* (HDW), performs a similar function.

Most of the inventory of chemicals and radionuclides at the Hanford Site now exists in facilities constructed on or in the vadose

---

[1]As noted in Chapter 4, the term *inventory* is used by the Integration Project to describe the quantities of radionuclides and chemicals that have been placed in storage and disposal facilities at the Hanford Site.

Inventory Technical Element

zone—in particular, the underground high-level waste tanks and waste ponds, pits, trenches, and cribs (see Chapter 2). Some of this inventory has migrated from these facilities into the vadose zone and groundwater. The need for characterization of these contaminant releases[2] in the vadose zone has been emphasized repeatedly in previous studies. For example, an earlier National Research Council (NRC) report stated that "an important component of a long-term commitment to remediating the single-shell tanks at the Hanford Site is an adequate understanding of the ... extent to which the soil and ground water beneath the tank farms have been contaminated. Characterization should continue until such an understanding has been obtained" (NRC, 1996, p. 28). A 1996 Department of Energy (DOE) review noted that "characterization of the vadose zone is an essential step toward understanding contamination of the groundwater, assessing the resulting health risks, and defining the concomitant groundwater monitoring program necessary to verify the risk assessments" (DOE, 1997b, p. P-3).

Under current plans for the Hanford Site, the majority of the current waste inventory in burial grounds and liquid disposal sites will be left in place (see discussion of the Cleanup of the Hanford Site in Chapter 2), as will past contaminant releases to the vadose zone and groundwater. Additionally, removal of waste from high-level tanks may result in further releases of contaminants to the subsurface (see NRC, 1996, p. 36-37). The estimation of long-term environmental impacts from the inventories and contaminant releases to be left in the ground requires an accurate knowledge of the amount of each contaminant in the soil (the *source term*), its chemical form (*speciation*, see Sidebar 5.1), and the rate at which each migrates through the subsurface, either in solution or in colloidal form (Sidebar 5.2). Assessment of source terms and migration rates, in turn, requires detailed characterization of the distribution of contamination in the environment as well as subsurface properties that control contaminant fate and transport. Since the Integration Project's science and technology (S&T) program mission is to aid in providing the data required for site decisions (see Chapter 3), characterization of the site must be one of its primary focuses.

The following are examples of decisions that will require some knowledge of waste inventories as well as past and possible future contaminant releases at the site:

1. *Disposition of existing waste sites in the 200 Area* (e.g. disposal cribs and canyons). Should such facilities be left in place

---

[2]The committee uses the term *contaminant release* to describe waste that has migrated out of disposal facilities and into the environment.

SIDEBAR 5.1 Chemical Speciation: Why it's Important

Chemical speciation refers to the chemical form of an element, ion, or molecule in a system. Processes that alter speciation include the transfer of electrons (oxidation-reduction), hydrolysis, the formation of chemical complexes between dissolved contaminant cations and neutral or negatively charged complexing agents (ligands), adsorption-desorption reactions at solid-solution interfaces, precipitation-dissolution, and biologically mediated reactions.

Chemical inventory data rarely include information on speciation. Yet the chemical form of an element often has a profound effect on its environmental behavior (e.g., mobility) and toxicity. Some elements have a relatively simple environmental chemistry: sodium, for example, exists primarily in ionic form (as Na^+) in aqueous environments. However, many of the contaminants of concern at Hanford exhibit complex speciation, and it is the environmental behavior of these species that must be considered in remediation planning.

Many of the contaminants at Hanford are classified broadly as heavy or transition metals. One property of these metals that distinguishes them from other contaminants is that they can exist in multiple oxidation states that are in thermodynamic equilibrium and can bond with a large number of compounds. The resulting complexes that are formed can vary in toxicity and mobility. In addition, many metals participate in oxidation-reduction reactions, and the oxidation state can substantially affect the element's ability to form the kinds of chemical complexes described above.

The following two examples are instructive. Cr(VI), a highly oxidized form of chromium, is genotoxic, mutagenic, and carcinogenic, and it tends to be mobile in groundwater due to its tendency to form soluble complexes. Cr(III), a less oxidized form, is essential for some enzyme activities and is less mobile in groundwater. Thus, the hazards posed by chromium in the environment can be evaluated only by knowing its speciation.

Plutonium (Pu) also exhibits a complex speciation and, like chromium, can exist in a variety of oxidation states under environmental conditions. Plutonium in certain oxidation states may form complexes with a wide range of ligands, including carbonate (CO_3^{-2}) and natural organic acids, and some of these complexes may be relatively soluble and mobile in groundwater. Thus, without a clear understanding of plutonium speciation, the prediction of its behavior in the environment is problematic (e.g., Kersting et al., 1999).

Inventory Technical Element

> The characteristics of environmental systems that govern chemical speciation can vary in time and space, particularly over the large spatial and temporal scales encountered at sites such as Hanford. Understanding the processes and conditions that lead to speciation transformations may allow scientists to better predict contaminant behavior over the wide range of environmental conditions found at Hanford, with improved confidence not possible by knowledge only of element mass concentrations.

essentially as is, or should additional steps be taken to reduce the potential for future contaminant migration?

2. *Retrieval of residual nonliquid waste[3] from single-shell tanks.* Would such retrieval result in substantial additional releases of contaminants to the subsurface, and would these releases pose a threat to the Columbia River or to other planned uses of the site?

3. *Disposition of tank farms.* If some residual waste is left in the tanks, will it pose a hazard to the river or other receptors? If so, what actions should be taken to minimize such hazards? For example, what benefits would be provided by surface barriers or other methods of infiltration reduction over the tank farms? When and where should such barriers be emplaced?

DOE has recognized the significance of the lack of characterization in the statement of needs for the Groundwater/ Vadose Zone Project: "Currently, information on contaminant distribution, physical association, and chemical form in the vadose zone ... is not adequate to forecast whether future breakthrough to groundwater will occur" (DOE, 2000a, p. B-5). DOE has also recognized the need for better characterization data: "This data set is needed as input to the [System Assessment Capability] SAC [to] allow the assessment of the cumulative effects of Hanford Site operations and remediation on the Columbia River and associated river-supported activities" (DOE, 2000a, p. B-74).

There is a substantial amount of characterization work now under way at Hanford, much of which is being conducted by the core projects (see Chapter 3). Existing dry wells[4] are being utilized for gamma-ray

---

[3] Especially solid waste attached to the sides and bottoms of the tanks that presumably will be removed by sluicing or other mechanical actions, which could damage the tanks (see Chapter 2).

[4] As noted in Chapter 2, wells completed in the vadose zone above the water table.

SIDEBAR 5.2 Do Colloids Transport Contaminants?

Colloids are collections of solid particles that range in size from approximately 1 nanometer to 1 micrometer (10^{-9} to 10^{-6} meters). Colloids include mineral particles and aggregations of organic compounds or mineral particles, biological entities such as viruses and bacteria, and organic macromolecules. Colloids are often proposed as transport for generally insoluble and, therefore, otherwise largely immobile substances. For example, colloids are proposed to have played an important role in the transport of plutonium, a highly insoluble element, in groundwater to a distance of some 1,300 meters from its assumed point of release at the Nevada Test Site (Kersting et al., 1999).

Although it is clear from experiment and field observation that colloids have the potential to travel substantial distances through saturated media, there have been no generally accepted reports of colloid-facilitated contaminant transport in field situations. Indeed, recent studies support the idea that the net effect of colloidal processes may, in some instances, be to retard rather than enhance the transport of strongly sorbed contaminants like cesium beneath the SX Tank Farm at Hanford. In a column study, simulated high-pH waste produced colloidal materials near the leading edge of the waste front (Wan et al., 2000). Rapid colloid generation, and attendant pore plugging and permeability reduction, had the net effect of retarding contaminant transport. Other studies have explored chemical mechanisms that can explain, at least in part, the unexpectedly deep migration of cesium through Hanford sediments without invoking colloids (Carroll et al., 2000). These studies showed that high sodium concentrations and highly alkaline conditions in simulated wastes greatly inhibited the sorption of cesium onto sediments, keeping cesium in solution and increasing its mobility.

An Environmental Management Science Program (EMSP) supported study also found that simulated high-pH waste immobilized native colloids in Hanford sediments on contact (Flury et al., 2000; Project Number 70135 in DOE, 2000a, Table 2-1, p. 2-4). Subsequent dilution of the waste, however, caused an "immediate release" of colloids from the sediments. Although colloids have not been observed in Hanford groundwater during Integration Project Investigations (PNNL, 2000b), fiscal year 1999 groundwater monitoring by core projects reported elevated levels of aluminum and iron as colloids in groundwater from well 299-W23-15 in the 200 Area (near the SX Tank Farm), elevated levels of uranium in an unfiltered groundwater sample from well 699-S6-E4A in the 300 Area (near the 618-10 burial ground

Inventory Technical Element   71

and 316-4 crib), and elevated levels of strontium-90 in unfiltered samples from well 399-3-11 in the 300 Area (Hartman et al., 2000). Many or all of these associations could be artifacts due to colloids generated at high pumping rates during sampling.

Although chemical speciation of contaminants (see Sidebar 5.1) is not determined during routine monitoring, the speciation of plutonium in Hanford groundwater was determined as part of an EMSP-supported study of actinide transport (Project Number 70132 in DOE, 2000a, Table 2-1, p. 2-9). This study showed that less than 6 percent of plutonium was bound to colloids in samples from four 100K Area wells (Buesseler et al., 2000). The project plans additional studies of groundwater in the 100-N Area and 200 East Area in 2001.

It is important to note that studies to date have been limited to colloids that were generated from interactions of simulated tank liquids with native Hanford sediments. Colloids generated during deliberate or inadvertent chemical precipitation within the tanks may differ with respect to transport behavior. S&T-supported research on the role of colloids in contaminant transport has been valuable thus far. Potential sluicing operations to recover solids from tanks could create, mobilize, and release colloidal contaminants. The issue of tank-formed colloids and their transport through the subsurface may be a fertile topic for S&T.

---

logging; push-in tools are being used for characterization at shallow levels; laboratory studies have been conducted on the chemistry of contaminants under specific conditions; and there is an ongoing effort to acquire additional characterization data by drilling new boreholes in the tank farms, as well as by geophysical logging of existing tank farm boreholes (DOE, 1999b). This work has provided a wealth of valuable information, which is to be compiled in field reports on individual tank farms. However, none of these planned tank farm field reports had been issued by DOE or its contractors by the time this report was being finalized for review in May 2001.

SCOPE OF INVENTORY TECHNICAL ELEMENT

The objective of the Inventory Technical Element is to develop understanding and models to estimate the following (DOE, 2000a, p. 1-4): (1) the partitioning of wastes in process streams that were discharged to waste disposal facilities in the vadose zone; (2) the behavior of specific contaminants in these waste streams; and (3) release mechanisms and rates from waste sites (e.g., burial grounds, liquid disposal cribs) to soils.

To address these objectives, S&T within this element is organized into six activities with 23 individual projects (Table 5.1):

1. *Unplanned releases.* The three projects under this activity (Inv-1 to Inv-3[5]) are focused on estimating the volumes and compositions of leakage that occurred from the high-level waste tanks. As noted in Chapter 2, documentation on these releases is quite limited. These projects were under way at the time of writing this report, and the information generated from the work is being supplied to the Office of River Protection's Tank Farm Vadose Zone Project (see Chapter 3).

2. *Soil site waste inventory.* The seven projects (Inv-4 to Inv-10) under this activity are intended to provide best estimates with associated uncertainties for contaminant source terms at various waste sites. Estimates are being made of releases resulting from early to recent site activities. These S&T projects were under way at the time of writing this report, and the information produced from these activities will be used by SAC (see Chapter 4) and the core projects.

3. *Models for selected contaminants.* The four projects (Inv-11 to Inv-14) under this activity are focused on modeling the distributions for technetium-99, tritium, and iodine-129 in Hanford waste streams as inputs to site-wide mass balance models. The stated intent of these projects is to generate and refine the inventory estimates for these radionuclides.

4. *Release models.* The two projects planned under this activity (Inv-15 to Inv-16) are intended to model contaminant releases from various solid waste burial sites and iodine-129 "scrubber saddles."[6] These models will be used by the SAC to predict future contaminant releases at these sites.

5. *River source term.* Four projects are planned under this activity (Inv-17 to Inv-20) to estimate the present-day releases of chromium, strontium-90, cobalt-60, and tritium to the Columbia River. These activities will be used by the SAC and the River Monitoring Project (see Chapter 3) and will also be used in the Columbia River conceptual model (see Chapter 8).

6. *Reconciliation of model and field data.* There are three projects planned under this activity (Inv-21-Inv-23), all of which will attempt to reconcile inventory estimates obtained from process models with field data from the soil sites. This activity will be repeated for each version of the SAC.

---

[5]The projects within this technical element are given these identification numbers in DOE (2000a; Table 4-1).
[6]Scrubber saddles are ceramic beds that were used to remove iodine-131 from fuel dissolver offgas in the chemical processing plants.

Inventory Technical Element

TABLE 5.1 Summary of S&T Activities and Planned S&T Projects Under the Inventory Technical Element

S&T Activity	S&T Projects Planned	Project Objectives	Project Duration (fiscal years)	Hanford Funding (thousand dollars)	EMSP Funding (thousand dollars)
Unplanned releases	3	Estimate volumes and waste compositions of unplanned releases from tanks containing three classes of waste: boiling waste, dilute waste, and concentrated waste	1999-2000	0^a	0
Soil site waste inventory	7	Provide a methodology and preliminary estimates of contaminant inventories for several types of intentional and unplanned discharges to soil in the 200 Area	1999-2001	710	0
Models for selected contaminants	4	Develop methodologies to describe the distribution of Tc-99, H-3, and I-129 in site waste streams	2000	190	0
Release models	2	Develop release models for residual contamination from various waste sites	2000-2001	160^b	0
River source term	4	Determine the inventories of Cr, Sr-90, Co-60, and H-3 released to the Columbia River	2001	195^b	0
Reconciliation of model and field data	3	Provide a reconciliation of field and model data for estimating releases to soil	2001-2003	390	0

NOTE: EMSP = Environmental Management Science Program
[a]The funding shown in the table will be provided by the Office of River Protection.
[b]Some or all of the funding shown in the table will be provided by the System Assessment Capability.
SOURCE: DOE, 2000a, Figure 4-1, Table 5-1.

EVALUATION OF WORK PLANNED UNDER THE INVENTORY TECHNICAL ELEMENT

There is not enough detail provided in the documentation of these projects, including the Integration Project Roadmap (DOE, 2000a), to undertake a detailed evaluation of the projects proposed or being conducted within this technical element. This review is therefore more general in nature, with only general comments offered on both work in progress and possible S&T gaps.

Integration Project staff described the S&T for this technical element during the committee's information-gathering sessions. They noted that the methods used to obtain estimates for contaminant inventories vary from waste stream to waste stream. Thus, one of the primary products of these projects will be documentation of the methods used to generate these inventory estimates. They noted that the methods and estimates were not intended for direct use in regulatory applications or remediation decisions, but rather were for use in the SAC and various core projects.

These staff acknowledged that because of the lack of adequate record keeping, especially during the early history of the Hanford Site, they expect to encounter future surprises regarding unexpected contaminants in the subsurface. They also emphasized that the most important issue is not the magnitude of the total inventory, but how much of that material actually poses a threat to the Columbia River and other potential receptors.

Among the major efforts under this technical element is the compilation of estimates of the characteristics of each waste stream at the site. The plutonium production process at Hanford consumed large quantities of uranium metal, acids, solvents, and other chemicals and produced waste streams containing dozens of radionuclides and chemical species. The quantities of uranium metal and chemicals used in processing operations can be estimated from procurement records, and the radionuclide and chemical outputs can be estimated from various process models. Much less well known, however, is the partitioning of chemicals and radionuclides into the large number of process streams and secondary waste streams during plutonium production and recovery.[7]

In the committee's judgment, the work under way in this technical element to obtain inventory estimates using process models is necessary

---

[7]For example, iodine-129 was partitioned into several process and waste streams during chemical processing of irradiated uranium slugs. Some was discharged to the atmosphere, some was captured in offgas scrubbers (see footnote 6), and some ended up in the high-level waste that was sent to the tanks and may later be transported into the environment through tank leaks.

to understand the current distribution of contaminants at the Hanford Site. It is not clear, however, whether such inventories can be estimated with sufficient confidence to be used in site-wide models such as the SAC without validation through field characterization studies. Moreover, although this process model work is essential, it is not sufficient to establish the current distribution of contaminant releases in the subsurface at the site. Very few measurements have been made of subsurface contaminant distributions, even though such measurements are essential for validating and reducing uncertainties in the process model estimates. At present, there are not sufficient data to establish either the distributions or the rates of migration of contaminants in the vadose zone. Data have been obtained from a few cores in the 200 Area, for example, and from a large number of gamma-ray measurements from shallow wells in the tank farms (see Figure 2.10). Although the shallow-well studies have provided valuable data on radionuclide distributions beneath some of the tank farms, contamination extends below or laterally to the wells in many cases. The Integration Project has acknowledged this problem, citing multiple instances in which contamination was found to extend as far as the bottom of these shallow wells in the AX, BX, BY, SX, TY, and U Tank Farms (DOE, 1998a, p. 4-66).

Similarly, modeling flow of fluids in both the vadose zone and groundwater requires a detailed knowledge of subsurface properties, especially hydrological parameters. Because of the size and complexity of the Hanford Site, obtaining these data by standard methods would be prohibitively expensive and time-consuming. The S&T work on methods to characterize contaminant distributions in the subsurface is also potentially applicable to subsurface property characterization.

Due at least partially to the high cost of drilling in soil with possible radioactive contamination, there has been very little coring in the 200 Area. Only a few "deep" wells (having depths between about 150 and 200 feet) have been drilled there, including a slant-drilled well that was completed recently in the SX Tank Farm. These efforts are yielding important data.[8] However, the site plans to drill only one additional borehole in other tank farms in each of the next two years. In view of the fact that there are 67 suspect "leaker" tanks and hundreds of waste disposal sites, the planned rate of characterization is not sufficient to establish, even approximately, the current distribution, speciation (Sidebar 5.1), or potential for transport (e.g., Sidebar 5.2) in the subsurface or important subsurface properties. This information is critical in evaluating the potential for future migration and in validating inventory estimates.

---

[8]The committee received a briefing on the SX Tank Farm slant borehole results at its March 2001 meeting. This work is still in progress, and the results have not yet been published.

The Integration Project has made a major effort to maximize the effectiveness of its characterization research by piggybacking on the activities of the core characterization projects.[9] Such efforts are highly commendable, but they are clearly insufficient to produce the detailed level of characterization data that will likely be needed to support remediation decision making at the site.[10]

Two options for achieving more rapid characterization of contaminant distributions and properties of the subsurface at the Hanford Site are (1) to increase the funds allocated to the characterization effort,[11] and/or (2) to develop and apply more cost-effective characterization methods. Due to the very high cost of drilling, retrieving, and analyzing core at contaminated sites,[12] it seems unlikely to the committee that sufficient funds can be made available to dramatically increase the rate of characterization using conventional methods. This suggests that investments to develop alternate methods are needed, particularly for characterization at depths greater than can be reached by push-in technologies[13] at Hanford.

The need to develop alternate characterization methods—in particular, minimally invasive technologies that work under a wide variety of ground conditions and allow real-time, in situ characterization—has been highlighted in another National Research Council report (NRC, 2000a). Such methods include steerable microdrills (drills having a diameter of a few centimeters) with downhole instrumentation for in situ measurements, and directional drills that allow samples to be obtained at long horizontal distances from the drilling site. **The committee agrees that this is an important need and recommends that development of cost-effective strategies and methods for characterization of contaminant distributions and subsurface properties of the vadose zone be made a priority of the S&T program.**

Since the development of cost-effective methods would likely find wide application across the DOE complex, much of the needed S&T work

---

[9]As noted in Chapter 3, the Integration Project refers to these piggybacking activities as "wrap-around science."
[10]The committee recognizes that it is not the responsibility of the Integration Project's S&T program to do subsurface characterization at the Hanford Site. Nevertheless, the committee believes that this characterization work must be done if site remediation decisions are to have sound technical and risk bases
[11]The high cost of characterization has long been an issue in the DOE complex (see GAO, 1992, 1998).
[12] DOE will spend about $2.65 million to drill, retrieve, and analyze core from the slant borehole in the SX Tank Farm (Mark Freshley, Pacific Northwest National Laboratory, written communication, May 8, 2001).
[13]Push-in technologies are generally useful for sampling the upper 30 meters or so of the subsurface, depending on ground conditions.

Inventory Technical Element

might be done in cooperation with other DOE programs—for example, the applied research and technology development programs sponsored by the Office of Science and Technology within the Office of Environmental Management, which has an annual R&D budget on the order of $200 million. The focus of S&T at Hanford might be to adapt and demonstrate technologies developed elsewhere to the needs and environmental conditions at the site.

S&T on subsurface properties and contaminant characterization is potentially transferable to monitoring development efforts (see Chapter 9). Therefore, the recommended characterization S&T, if planned carefully, could also improve subsurface monitoring capabilities. Consider, for example, the use of characterization boreholes for monitoring. Current practices, which are driven largely by regulations, often result in the permanent plugging of characterization boreholes after characterization is completed to prevent the future spread of contamination. Once plugged, these boreholes cannot be used for monitoring. The development of methods to develop characterization boreholes that do not have to be permanently plugged to prevent contaminant spread could advance monitoring capabilities at Hanford and other DOE sites.

In addition to radionuclide contamination, the vadose zone and groundwater in the 200 West Area are also contaminated with hazardous chemicals. As discussed in Chapter 2, for example, large quantities of carbon tetrachloride (as dense nonaqueous phase liquid [DNAPL]) were discharged to cribs in the 200 Area between 1955 and 1973, and most of this contamination is estimated to remain in the subsurface (DOE, 2000e). DOE has been unable to locate the source of this contamination and does not know whether it poses a long-term threat to the river.

The amounts and locations of carbon tetrachloride in the vadose zone and groundwater are important and unresolved issues. The selection of remediation options and the effectiveness of recharge controls to keep the contamination from spreading depend to a great extent on the location of contaminant source terms in the subsurface. The characterization of DNAPL bodies in the subsurface, especially the vadose zone, is a difficult technical challenge. Developing methods to obtain such information is an appropriate S&T program task.

The Remediation Technical Element is working on the carbon tetrachloride plumes in the 200 West Area to assist in the development of a strategy for corrective actions. As discussed in Chapter 9, however, all of this work is being supported through the Environmental Management Science Program, and none of it appears to be focused directly on delineating the locations of DNAPL in the subsurface. **The committee recommends that the S&T program develop a plan to characterize carbon tetrachloride contamination in the 200 West Area, including a**

plan to detect the existence of pure phases in the groundwater and vadose zone. This plan could be used by the core programs (see Chapter 3) to do the actual characterization work.

6
Vadose Zone Technical Element

The Integration Project's Vadose Zone Technical Element supports studies to obtain a better understanding of contaminant behavior in the unsaturated zone at the Hanford Site and to develop conceptual models, numerical models, and parameter databases for the System Assessment Capability (SAC; see Chapter 4). The vadose zone is arguably the most important region of the Hanford Site from both a scientific and an environmental restoration perspective: it contains most of the chemical and radionuclide contaminants that have been discharged or leaked into the environment and is host to the site's waste storage and disposal facilities, including the high-level waste tanks, burial pits and trenches, disposal ponds and cribs, and injection (or "reverse") wells (Chapter 2). The present-day distributions and chemical forms of contaminants in the vadose zone are poorly known, as are the fate and transport processes that will govern the future migration of these contaminants to the groundwater and the Columbia River.

This chapter provides a brief review and assessment of the work supported under this technical element. The main sources of information used in this assessment are the Integration Project Roadmap (DOE 2000a), other DOE documents (DOE 1999e, 2000g), and briefings received during the committee's information-gathering meetings. It was apparent to the committee from these briefings that the Vadose Zone Technical Element is still in the early stages of development and that the schedule for S&T work is in flux owing mainly to budget reductions (Chapter 10).

THE VADOSE ZONE: WHAT IS IT, AND WHY IS IT POORLY UNDERSTOOD?

The *vadose zone*, also called the *unsaturated zone*,[1] is that portion of the earth's crust between the land surface and the water table. It includes the capillary fringe (a region above the water table that

---

[1] The adjective *vadose*, from the Greek word "shallow," was introduced by Posepny in 1894 to designate water in the unsaturated zone, although subsequent usage included shallow groundwater as well (Meinzer and Wenzel, 1942). In recent years, however, the term *vadose zone* has been used more or less synonymously with unsaturated zone, and the committee uses these two terms interchangeably in this report.

contains water held by capillary action), perched water bodies, and other features that may be temporarily or permanently filled with water.

The unsaturated zone contains solids, liquids, and gases. The *solid phase* consists of rock and mineral particles interspersed with organic solids, as well as plant and animal life. Solid particles vary in size from fractions of a micron in clays to millimeters in sands and gravels. The largest particles can be meters across and have substantial internal porosity.

The *liquid phase* is composed primarily of aqueous solutions that exhibit variable concentrations throughout the vadose zone. The distribution of solutes varies within individual pores due to electrical and chemical gradients at liquid-solid and liquid-gas interfaces. At sites such as Hanford, the liquid phase contains a variety of inorganic and organic contaminants, including nonaqueous liquids. Some, such as alcohol, mix completely with water. Others, such as carbon tetrachloride, form distinct phases known as nonaqueous phase liquids (NAPLs).

The *gas phase* is generally similar in composition to that of the above-ground atmosphere, except for elevated concentrations of water vapor and carbon dioxide. At contaminated sites like Hanford, the vadose zone gas phase contains semivolatile and volatile organic compounds as well. Gas transport is driven by compositional, pressure, and thermal gradients. Near sources of contamination, this transport is difficult to model.[2]

The vadose zone typically contains from 20 to 50 percent porosity by volume. Pores have irregular shapes and complex interconnections that elude precise description. Pores in sediment arise from depositional features that are modified by postdepositional processes. Modification of pores occurs during soil formation, weathering, and biological processes. Consequently, pore geometries tend to be spatially heterogeneous and anisotropic. In addition to the interconnected pore space between grains, passageways for fluids include burrows, root channels, fractures, and human artifacts including well bores and corroded pipes.

Small particles (e.g., clay minerals) may contain large amounts of porosity and surface area, up to hundreds of square meters per gram. Surfaces of wetted clays are electrically charged and interact with charged species in the liquid phase. At Hanford, electrochemical interactions were assumed to bind certain contaminants, particularly cesium, strongly to the solid phase, retarding their migration (see Chapter 1).

Unsaturated zones are chemical and mechanical systems in disequilibrium in which fluids and solutes move in response to gradients in

---

[2]Neither diffusion theory nor advection theory alone accurately predicts gas transport near sources of contamination.

free energy. Transport rates are approximately proportional to gradients in free energy, but the proportionalities are nonlinear functions of saturation (Figure 6.1). Because pores contain varying amounts of gas and liquid, transport parameters are represented by saturation-dependent functions rather than by constant values as in the saturated zone. Additionally, some transport parameters exhibit hysteresis as a function of saturation— that is, they have different values depending on whether the system is being wetted or dried.

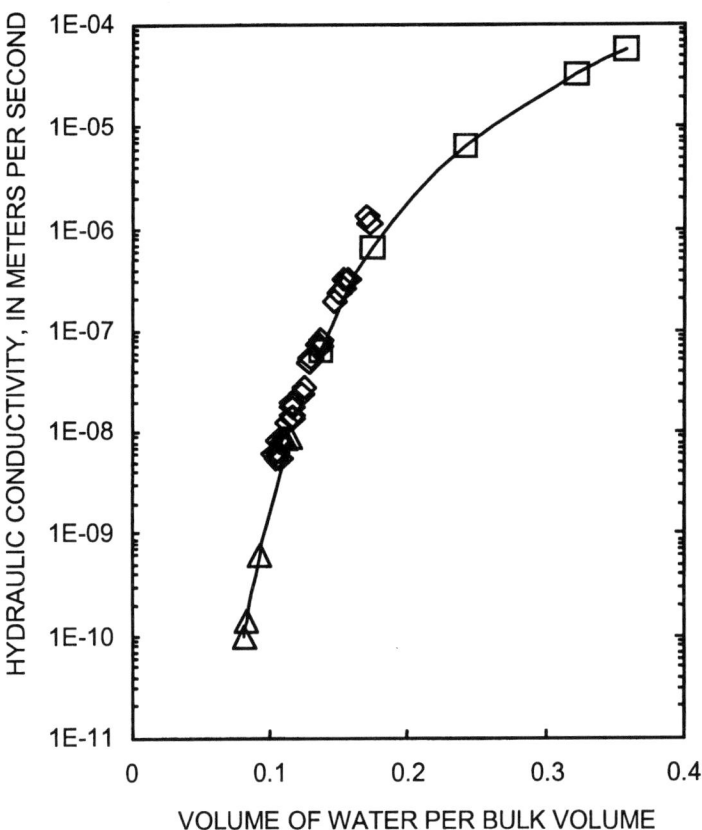

Figure 6.1. Dependence of unsaturated hydraulic conductivity on saturation, expressed by the volume of water per bulk volume, in sand. Different methods of determination (squares, diamonds, and triangles) are necessary to span the range of interest accurately. SOURCE: Stonestrom, 1996.

These nonlinearities make it difficult to obtain representative measurements and tend to amplify modeling errors. A 5 to 10 percent uncertainty in saturation, for example, can lead to an order-of-magnitude uncertainty in predicted transport rates. Standard methods for hydraulic conductivity determination are limited to one or two orders of magnitude in range and become impractical at low water content. Hydraulic conductivities are often inferred but rarely measured at saturations less than 50 percent.

Quantifying water and solute movement through the vadose zone is particularly difficult in arid regions. Most precipitation reaching the ground returns to the atmosphere through evapotranspiration; groundwater recharge is thus the difference between two nearly equal quantities. The amount of water crossing the land surface as liquid or vapor depends on dynamic meteorological and plant conditions that change by the hour. Evapotranspiration is therefore difficult to measure and model. Because of this, water-balance estimates of recharge are subject to large errors. These uncertainties are amplified by climate change, which can alter flora and fauna and produce major shifts in recharge locations and amounts.

In summary, the vadose zone is a complex system of interacting physical, chemical, and biological processes. Mathematical models of transport incorporate parametric functions that exhibit nonlinearity and hysteresis, complicating hydrogeological characterization. Heterogeneities exist at scales from individual mineral grains to geologic formations, further complicating characterization. For all of these reasons, modeling the fate and transport of contaminants through the vadose zone presents a difficult technical challenge.

SCOPE OF VADOSE ZONE TECHNICAL ELEMENT

The Vadose Zone Technical Element comprises five broad science and technology (S&T) activities and, within these, 27 individual "projects" (Table 6.1):

1. *Field investigations of representative sites:* This activity includes six projects to develop an improved understanding of contaminant distributions beneath selected tank farms and at 200 Area soil waste sites.[3]

---

[3]Waste sites (e.g., tanks, ponds, cribs, trenches, landfills) in the 200 Area have been grouped based on waste inventories (DOE, 1997c), and efforts are under way to characterize representative sites from each of these groups. These sites are referred to by the Department of Energy as 200 Area soil waste sites.

Vadose Zone Technical Element 83

2. *Transport modeling:* This activity includes eight projects to obtain an improved understanding of fate and transport processes beneath selected tank farms and at 200 Area soil waste sites.

3. *Waste and sediment experiments and models:* This activity includes six projects to obtain kinetic and thermodynamic data on key contaminants to determine first-order hydrochemical reactions controlling contaminant behavior in sediments beneath tank farms and at 200 Area soil waste sites.

4. *Vadose zone transport field studies:* This activity includes four projects to develop an improved understanding of water and solute movement, reactive transport, and migration pathways in vadose zone sediments.

5. *Advanced vadose zone characterization*: This activity includes three projects on advanced characterization technologies to support the vadose zone transport field studies in the 200 East Area and 200 West Area and to evaluate tools for monitoring contaminant plumes in the vadose zone beneath tank farms.

As shown in Table 6.1, work on projects under the Vadose Zone Technical Element is planned to run from fiscal year 1999 through fiscal year 2004, and some of the early work was being completed as the committee finished its information gathering for this report. The total planned funding for this technical element is about $42.6 million, of which $17.8 million is being provided to Environmental Management Science Program (EMSP) projects from the fiscal year 1999 competition.[4] The actual budgets for the Vadose Zone Technical Element have been lower than indicated in Table 6.1 owing to funding cutbacks (see Chapter 10).

EVALUATION OF WORK PLANNED UNDER THE VADOSE ZONE TECHNICAL ELEMENT

As of early 2001, most of the technical work to be done within the Vadose Zone Technical Element either had not been started or was not yet completed. Consequently, there is little scientific or technical output in the form of peer-reviewed reports or papers available for the committee's evaluation. The committee has therefore focused its efforts on reviewing the written plans for this work and providing responses to the following five questions that were developed to address the statement of task for this study (Chapter 1):

---

[4]The science program projects are under way and are scheduled to be completed in fiscal year 2003.

TABLE 6.1 Summary of S&T Activities and Planned S&T Projects Under the Vadose Zone Technical Element

S&T Activity	S&T Projects Planned	Project Objectives	Project Duration (fiscal years)	Hanford Funding (thousand dollars)	EMSP Funding (thousand dollars)
Field investigations of representative sites	6	Develop an improved understanding of contaminant distributions beneath selected tank farms and at 200 Area soil waste sites	2001-2004	7,830	1,600
Transport modeling	8	Obtain an improved understanding of fate and transport processes beneath selected tank farms and at 200 Area soil waste sites	2001-2004	3,840	600
Waste and sediment experiments and models	6	Obtain kinetic and thermodynamic data on key contaminants to determine first-order hydrochemical reactions controlling contaminant behavior in sediments beneath tank farms and at representative 200 Area soil waste sites	2001-2004	3,500	8,000
Vadose zone transport field studies	4	Develop an improved understanding of water and solute movement, reactive transport, and migration pathways in vadose zone sediments in the 200 East Area and 200 West Area	1999-2004	8,120	3,900
Advanced vadose zone characterization	3	Use advanced characterization technologies to support the vadose zone transport field studies in the 200 East Area and 200 West Area, and evaluate tools for monitoring contaminant plumes in the vadose zone beneath tank farms.	2000-2003	1,500	3,700

NOTE: EMSP = Environmental Management Science Program
SOURCE: DOE, 2000a, Figure 4-1, Table 5-1.

Vadose Zone Technical Element

1. Can the objectives of the planned work be achieved?
2. Does the planned work represent new science?
3. Can the planned work have an impact on cleanup decisions at the Hanford Site?
4. Does the planned work address the important issues?
5. Are there other concerns, comments, or suggestions that should be considered by the Integration Project in executing the planned work?

The five S&T activities are described and evaluated in the following sections. More written documentation is available for some projects in this technical element than in the Inventory Technical Element (Chapter 5). Consequently, the committee is able to provide a more detailed review.

Field Investigations of Representative Sites

Six separate projects are planned under this activity to improve understanding of contaminant distributions in the vadose zone in the 200 Area. These projects are designed around field investigations at what the Integration Project calls "representative sites," that is, sites designed by the Integration Project to be broadly representative of the population of waste sites that exist in the 200 Area based on characteristics such as waste type and vadose zone geology.

The scale of evaluation for most of the projects under this activity is the individual mineral, although studies of intact cores and homogenized core material will be undertaken to examine questions related to contaminant migration. Three specific processes and/or attributes of waste-soil interactions will be examined for (1) the potential for immobilization of technetium and cesium; (2) the influence of temperature; and (3) aluminum activity on subsurface mobility of waste constituents. Other task objectives are more open-ended.

Two projects (VZ-1[5] and VZ-3) are focused on understanding chemical and hydrochemical processes beneath leaking single-shell tanks in the S-SX Tank Farm, which contain highly concentrated waste from the PUREX (Plutonium-Uranium Extraction) process, and the B-BX-BY Tank Farm, which contains dilute high-level waste from other chemical processing operations. Some of this work is being conducted in cooperation with the Office of River Protection, which is drilling wells in the tank farms to obtain contaminated core samples from beneath tanks that are suspected to have leaked.

---

[5]The projects under each of the six activities are given these identification numbers in DOE (2000a, Table 4-1).

One project (VZ-2) is focused on understanding chemical and hydrochemical processes beneath other 200 Area soil waste sites, especially sites that received significant inventories of technetium, actinides, and dense non aqueous phase liquid (DNAPLs). Some of this work also is being conducted in cooperation with Hanford core projects (Chapter 3).

Three projects (VZ-4, VZ-5, and VZ-6) are focused on developing conceptual models of the important processes controlling contaminant distributions beneath leaking single-shell tanks and soil waste sites in the 200 Area. This information will serve as input to future revisions of the System Assessment Capability (see Chapter 4).

Can the objectives of the planned work be achieved?

A significant portion of the proposed tasks involves characterization of contaminant-sediment associations. Presumably, once such associations are elucidated, the development of hypotheses regarding the mechanism of interaction will follow. An objective of the conceptual model development is to obtain a comprehensive understanding of the important processes controlling contaminant distribution beneath waste tanks. Criteria for the successful completion of the tasks are unclear, and the level of understanding required to meet data needs is not defined.

Does the planned work represent new science?

The scientific merit of the proposed characterization work appears to be good, particularly with the application of state-of-the-art analytical techniques such as x-ray absorption spectroscopy. Experience gained from working on Hanford Site materials should be applicable to contaminant-sediment interaction questions at other Department of Energy (DOE) sites.

Can the planned work have an impact on cleanup decisions at the Hanford Site?

The proposed work will focus on materials of specific concern to the Hanford Site. However, an important question remains to be answered: Is the scale of analysis appropriate for the scale at which site decisions must be made? A goal of this work is the incorporation of conceptual models into the SAC Rev. 3 (see Chapter 4), but the S&T program has not demonstrated how mineral-scale studies will fit into a site-wide simulation model such as SAC. Translating the information derived from mineral-grain studies up to the spatial scales represented by

Vadose Zone Technical Element

site-wide models like the SAC is not a trivial task (see Sidebar 6.1). This is especially true for subsurface structure, where the challenge is to understand the dominant components of heterogeneity at large scales.

Does the planned work address the important issues?

The Integration Project has not provided an explicit link between the planned work and the issues to be addressed, especially within the context of future decisions to be made at the site. Nor have the data quality objectives or criteria for success been determined.

Developing an understanding of mechanisms of contaminant-sediment interaction is important for providing confidence that reactive transport models are conceptually correct. However, the scope of the problem and site heterogeneity will confound efforts to achieve closure on the physicochemical controls on contaminant migration. It is not clear how the importance of a process will be determined. In short, how is it possible to make conclusions regarding the relevance of processes to site needs from 500-gram sediment samples (DOE, 2000g, p. 5.4)? What will be the criteria for selecting samples for detailed analyses?

Are there other concerns, comments, or suggestions that should be considered by the Integration Project in executing the planned work?

The committee is concerned about how the S&T program intends to set the data quality objectives for supporting sound management decisions. Data quality objectives include the type and distribution of data (e.g., What are the cesium concentrations in the vadose zone at an appropriate spatial distribution and sampling density?) and uncertainty requirements (i.e., How well does a particular parameter value need to be known?).

The data quality needs can be considered only in the context of a specific management tool (e.g., SAC), because not all data will be critical to uncertainty reduction. For example, it may be necessary to know the value of the sorption parameter only to within an order of magnitude in a particular system to estimate a particular risk component. However, the level of certainty (precision) that has to be achieved cannot be defined in the absence of identifying the specific need.

Transport Modeling

According to the Integration Project Roadmap (DOE 2000a), eight separate projects are planned to improve understanding of fate and transport processes in the vadose zone in the 200 Area. Three projects

SIDEBAR 6.1 The "Scaling" Challenge at Hanford

The process of using observations made at one set of spatial and temporal scales to understand processes or postulate behaviors at another set of scales is commonly referred to as *scaling*. The scaling issue is confronted by scientists and engineers in a wide range of technical endeavors. At Hanford, the scaling issue comes into play when using "contemporary" scientific data to understand and predict long-term, site-wide contaminant fate and transport behavior. Scaling is also confronted when results of laboratory-scale experiments are extended to explain field-scale observations.

Site cleanup and waste management decisions at Hanford will be made with the benefit of site-scale models such as the SAC (see Chapter 4) that are based on scientific data collected over a relatively narrow range of spatial and temporal scales. The most complete environmental data sets at Hanford have been accumulated for only about 50 years, and many for much shorter times. Most of the scientific data that can be used to parameterize the SAC—much of it produced by the Integration Project S&T program—is based on laboratory work and small-scale field studies. These data sets may not reflect the full range of characteristics and properties that are important to contaminant transport processes at site scales, for example, those from extreme events (see Sidebar 9.1), which become increasingly important as the time horizon is extended.

The range in spatial scales of concern at Hanford spans more than 15 orders of magnitude (see Figure 6.2; see also Appendix C). Work at the molecular level (10^{-10} meter) that is part of the S&T program has as its goal understanding basic physical-chemical properties that affect contaminant fate and transport processes, in part to provide confidence in the processes embodied in site-wide models such as the SAC. These will be used to model site-wide contaminant plumes (10^4 meters or more in width) that could potentially impact large stretches of the Columbia River (10^5 meter or more in length).

The coupling of space and time scales at Hanford poses unusual challenges. Site impacts must be considered in terms of both the rate of the physical, chemical, and biological processes that transform contaminants (see Sidebar 5.1), and the rate at which physical processes act to transport contaminants and their breakdown products. For example, sediment transport in the Columbia River may be rapid relative to sorptive processes that cause suspended particles to "scavenge" certain contaminants from the dissolved phase. Thus, sorptive processes may be unimportant compared to the physical transport of dissolved forms of the contaminant. In contrast, if the

Vadose Zone Technical Element

> sorption of contaminants by groundwater aquifer solids is rapid compared to the rate of groundwater flow, the retardation of contaminants may be described by equilibrium chemical models. Otherwise, complex kinetic models must be used to predict system behavior accurately.
>
> It is not always clear how to scale-up contaminant behavior, but the outlook is by no means bleak. Hydrologists have, for example, achieved success in understanding the influence of scale on some transport parameters (Appendix C). In some cases, it may be possible to understand contaminant behavior on long time scales through the observation of natural analogues—for example, understanding controls on uranium transport through an examination of the processes that lead to the formation of uranium deposits. However, many of the contaminants at Hanford have no obvious natural analogues. The challenge, then, is to predict system evolution at spatial and temporal scales for which there is no environmental analogue or opportunity to fully test the outcome and accuracy of predictive models.

(VZ-07, VZ-09, VZ-10) target selected waste management areas containing single-shell tanks (S-SX Tank Farm, B-BX-BY Tank Farm, T-TX-TY Tank Farm), and one project (VZ-08) targets "high-priority" but unspecified 200 Area soil waste sites. All four projects have as their scope "preliminary evaluation of key transport processes affecting contaminant transport" (DOE, 2000a, Table 4-1).

One project (VZ-11) will provide the SAC Rev. 2 with "evaluations of key contaminant transport processes beneath SSTs [single-shell tanks]." Another project (VZ-12) will provide the SAC Rev. 3 with evaluations of "coupled fluid flow and multicomponent reactive transport" (DOE, 2000a, Table 4-1).

Two projects (VZ-13, VZ-14) will provide modeling support for the experimental design of field-scale infiltration and reactive tracer experiments. The first will be carried out at an uncontaminated site in the 200 East Area (the site of the current vadose zone field transport experiment). The second will be carried out at an uncontaminated site in the 200 West Area that is yet to be selected.

Each of these eight projects is to produce "a documented suite of process models and simulation results" for the targeted area (DOE, 2000a, Table 4-1). In addition, one EMSP project is linked to this activity. That project is entitled "Quantifying Vadose Zone Flow and Transport Uncertainties Using a Unified, Hierarchical Approach." Its purpose is to develop "a general approach for modeling flow and transport in a heterogeneous vadose zone using geostatistical analysis, media scaling, and conditional simulation to estimate soil hydraulic parameters at unsampled locations from field-measured water content data and a set of scale-mean hydraulic parameters" (DOE, 2000a, Table 2-1).

Can the objectives of the planned work be achieved?

It appears likely that objectives will be achieved only to a limited extent. An important limiting factor is the lack of data for calibration and testing purposes. There is little information on the three-dimensional distribution of contaminants under the tanks, and almost no data exist on conditions in the deep (>30 meters) vadose zone. Data to assess lateral movement of contaminants from tank areas are largely unavailable.

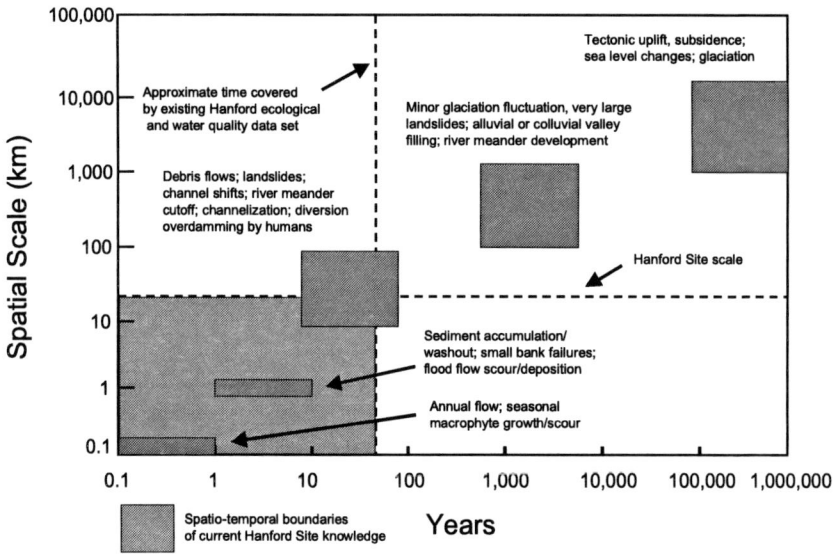

Figure 6.2 Spatial and temporal scales of geologic and hydrologic processes relative to the Hanford Site process data set. SOURCE: Adapted from Frissell et al., 1986.

Vadose Zone Technical Element

Does the planned work represent new science?

Although modeling reactive transport through heterogeneous sediments is not new, several features of the targeted systems are unique to the Hanford Site. Unique features include substantial contaminant chemistries (high pH, high ionic strength, and unusual compositions) and thermal effects. Hanford-specific features also include high levels of sediment heterogeneity in the glacial-lake outburst flood deposits that underlie the site.

Can the planned work have an impact on cleanup decisions at the Hanford Site?

Well-constructed and calibrated transport codes could directly inform remediation and stewardship decisions at the site, especially if their use has been formally linked to specific site decisions. Defensible transport codes should help form the scientific and technical bases for cleanup. Even if transport codes are not sufficiently accurate or well constrained for management decisions, an important role of modeling activities can be to identify important gaps in understanding.

Does the planned work address the important issues?

The Integration Project has not provided an explicit link between the planned work and the issues to be addressed, but the committee sees several such links. The planned work could help explain the cesium "anomaly" discovered in the deep vadose zone beneath one of the tank farms (Chapter 1). In fact, the transport-modeling activity emphasizes the tank farms, where pressing decisions on tank waste retrieval and tank closure loom (Chapter 2). The activity also emphasizes modeling to support future revisions of SAC. Given the inability of models that have been used in the past to predict the observed migration of contaminants, the transport-modeling projects address some of the most important issues at Hanford. It is also likely, however, that the cribs, ponds, tile drains, and plutonium production canyons are not receiving the S&T attention they require. Also, there are potentially significant S&T gaps related to modeling microbiological and sediment transport processes.

Are there other concerns, comments, or suggestions that should be considered by the Integration Project in executing the planned work?

This review was based on descriptions of S&T activities in the Integration Project Roadmap (DOE 2000a) and other DOE documents (DOE 1999e, 2000g). The committee was unable to offer detailed critiques of these S&T activities because documentation of objectives and work plans were either missing or quite limited. Nevertheless, one of the main concerns that emerges from this review is that with the exception of the targeted tank farms, most vadose zone waste sites (the 200 Area soil waste sites) are not being studied, which leaves a large knowledge gap. Given the large number of waste management areas at Hanford, the S&T effort seems to be spread thin. It is worth noting that the schedule in DOE (2000, Figure 4-1) has slipped and that fiscal year 2000 funding for the transport-modeling activity is one-third of the planned amount (DOE, 2000g, p. 5-1). One aspect of the S&T program that is yielding very positive results is the solicitation and active involvement of modeling talent from other DOE national laboratories (Los Alamos National Laboratory, Lawrence Berkeley National Laboratory, and Lawrence Livermore National Laboratory).

It is unlikely that any modeling effort will provide usable results unless there are appropriate data for calibration and verification. There will also have to be formal procedures for comparing the field observations with the modeled predictions in view of the substantial uncertainties expected in both. Acquisition of such data must be an integral part of the S&T plan. The Integration Project is directing some of the needed data collection, but the committee believes that such efforts must be increased.

Waste and Sediment Experiments and Models

The stated goals of this activity are (1) to develop an improved understanding of key geochemical phenomena in target waste sites by conducting kinetic and thermodynamic studies of contaminants of concern using uncontaminated and contaminated sediments to determine proximal chemical and hydrochemical reactions and (2) to use the data from the first goal in the development of numerical models for describing contaminant transport through unsaturated columns. Six projects are planned under this activity to meet these objectives. Four of the projects (VZ-15, VZ-16, VZ-17, VZ-18) will involve kinetic and thermodynamic studies to understand hydrochemical reactions beneath the S-SX Tank Farm, B-BX-BY Tank Farm, T-TX-TY Tank Farm, and as yet unspecified 200 Area soil waste sites. The remaining two projects (VZ-19, VZ-20) will focus on the development of numerical models that describe those reactions.

Vadose Zone Technical Element

Can the objectives of the planned work be achieved?

The goals of these projects are rather open-ended—for example, "an improved understanding" and "to develop data." The data quality objectives are not specified.

Does the planned work represent new science?

Because much of the work will focus on Hanford Site materials, the work will be new. It is unclear, however, whether the laboratory experiments will be new in approach and whether the questions addressed, and the modeling techniques employed, will be new in a generic sense (i.e., an advance in the science).

Can the planned work have an impact on cleanup decisions at the Hanford Site?

Because there is not a clear link between the technical element activities and specific management decisions, the applicability of these tasks to cleanup decisions is not evident. No specific hypotheses are listed for testing, and the tasks give one the impression that they are meant to characterize system attributes rather than address testable hypotheses.

Does the planned work address the important issues?

The task descriptions provided to the committee are not sufficiently defined for it to ascertain the central issues to be resolved.

Are there other concerns, comments, or suggestions that should be considered by the Integration Project in executing the planned work?

As with the field investigation tasks, the problem remains of how to set the data quality objectives for supporting management decisions. A key aspect is the accurate characterization and modeling of chemical speciation and transformations in time and space.

Vadose Zone Transport Field Studies

The stated primary objective of these field studies, to be conducted at uncontaminated sites, is to collect data sets to verify conceptual and numerical models that describe transport through the vadose zone. A secondary objective is to test advanced characterization techniques at the Hanford Site under controlled conditions. Some key science issues driving these studies are plume identification and delineation, upscaling techniques (Sidebar 6.1), effects of elevated salt concentration, and preferred pathway analysis. The additional science issues of thermal and accelerated recharge effects on contaminant migration from tank leaks and colloidal transport in coarse heterogeneous sediments are identified in the fiscal year 2000 work plan (DOE, 1999e) but not in the fiscal year 2001 plan (DOE, 2000g).

In fiscal year 2001, a new field experiment with high-salt-concentration tracers is to be conducted at the existing test facility in the 200 East Area, the so-called "Sisson and Lu site," which consists of a concentric array of wells around a central injection well. Plans for a new field testing facility, tentatively in the 200 West Area, are to be developed starting in fiscal year 2002. The crucial issue of upscaling methodologies is deferred to a workshop in fiscal year 2002 as part of the development of a test plan for the deep (>20 feet)[6] vadose zone transport studies in the 200 West Area.

Reactive transport field experiments (VZ-22 and VZ-24) are identified prominently in the project descriptions given in Table 4-1 in the Integration Project Roadmap (DOE 2000a), but aside from the high salt concentration reactive transport experiment, they do not seem to be a major consideration in the detailed work plans (DOE, 1999e, 2000g). The issue of field-scale reactive transport is largely unresolved and would seem to be central to many problems at Hanford. Overall, the field investigations are intended to integrate with the field investigations of representative field sites and the transport-modeling activities, and are to provide results that will be used by the SAC and the Office of River Protection project for model verification tests.

Can the objectives of the planned work be achieved?

Clearly, many objectives of the proposed field experiment are achievable, but it is not clear that the resulting data collected will be adequate to definitively resolve the scientific issues identified. The approach to be taken to the difficult issues of upscaling and preferential pathways is not clear from the available documentation, and the efforts directed to these issues are deferred until late in the project. Field experimentation of this kind is very important from both basic and applied perspectives, but it is generally difficult to anticipate the outcome of such efforts.

---

[6]The description of what constitutes the "deep" vadose zone is different in various documents reviewed by the committee.

Vadose Zone Technical Element

Does the planned work represent new science?

Field experimentation of this kind is unique, particularly for the deep, dry, unusually heterogeneous vadose zone at Hanford. If these experiments can be used to establish an effective approach to characterize and simulate such large-scale heterogeneous nonlinear systems, this would be a major scientific contribution.

Can the planned work have an impact on cleanup decisions at the Hanford Site?

Certainly field-tested and validated techniques for predicting large-scale, long-term fate and transport of contaminants in the vadose zone at Hanford would be useful in cleanup decisions, both for individual contaminated sites and for a site-wide effort such as the SAC.

Does the planned work address the important issues?

Field experiments of this kind, if adequately designed and executed, are central to efforts to reliably assess the fate and transport of contaminants currently in the vadose zone at Hanford and to predict the behavior of wastes that may be deposited in the vadose zone in the future. Improved characterization techniques for both contamination and media properties are also very important.

Are there other concerns, comments, or suggestions that should be considered by the Integration Project in executing the planned work?

A major concern is the lack of emphasis on upscaling techniques early in the effort. If experiments of this kind are designed around specific upscaling techniques from the very beginning, it is much more likely that the necessary and sufficient data will be collected and definitive conclusions will evolve. Another concern is the unrealistic time frame for the completion of these experiments. The processes involved are very slow, particularly for the deep vadose zone experiments with reactive transport, whose initiation of which is deferred until late in the project. It is unrealistic to suggest, as implied by Figure 4-1 of DOE (2000a), that meaningful field experiments of this kind can be completed by early fiscal year 2003.

Advanced Vadose Zone Characterization

Three projects (VZ-25, VZ-26, VZ-27) are planned under this activity. Two involve field tests of characterization technologies for delineating moisture and contaminant plumes at the vadose zone field transport study sites in the 200 East Area and 200 West Area (VZ-25 and VZ-26). The third project (VZ-27) will evaluate characterization tools to support single-shell tank retrieval and closure decisions (VZ-27). Characterization tools being evaluated include tracers, tensiometers, neutron-logging devices, pore-water monitoring devices, cone penetrometers, and geophysical imaging techniques.

Outcomes of the first two projects will be documented tests that describe the performance of the characterization techniques in the field-scale transport studies. The outcome of the third project will be an evaluation of tools for delineating plumes of non-gamma emitting contaminants such as technetium-99.

In addition, five EMSP projects are linked to this activity. Two deal with developing sensors for technetium and organochlorides. The other three deal with geophysical techniques for characterizing flow and transport in the vadose zone.

Can the objectives of the planned work be achieved?

This is an area is which investments in S&T could yield high returns. Techniques for characterizing the shallow vadose zone (i.e., from 0 to about 15 meters in depth) have already been evaluated in field tests that started in May 2000. Characterization of the deeper vadose zone still appears problematic. Surface-based geophysical techniques lose resolution with depth. Subsurface techniques are limited by access limitations and concerns about creating pathways for preferential flow. It is unclear, however, whether characterization objectives for the deep vadose zone can be achieved.

Does the planned work represent new science?

The limited scope of the projects under this activity supports relatively little development of new or Hanford-specific techniques. Rather, advanced techniques developed at other sites are being tested and evaluated in Hanford sediments. Several of the techniques being evaluated represent emerging scientific advances.

Vadose Zone Technical Element

Can the planned work have an impact on cleanup decisions at the Hanford Site?

There is wide agreement that lack of vadose zone characterization hampers remediation decisions. To a large degree, lack of vadose zone characterization reflects limitations of available techniques. Advances in characterization technology will significantly support cleanup decisions. This point is discussed further in Chapter 5.

Does the planned work address the important issues?

All of the parameters currently being targeted are important. The issue of deep characterization and monitoring needs more attention. Tensiometers (used for measuring pore-water pressure) are limited to relatively moist conditions; generalized measurement of pore-liquid potential has to be addressed. The lack of techniques for measuring pore-liquid chemistry appears to be a significant gap. Techniques for thermal and microbiological characterization also appear to have gaps.

Are there other concerns, comments, or suggestions that should be considered by the Integration Project in executing the planned work?

This review is based on brief descriptions of S&T activities in the Integration Project Roadmap (DOE 2000a) and other DOE documents (DOE 1999e, 2000g) and is limited in breadth and depth because activity descriptions are lacking in detail. The advanced vadose zone characterization technical element appears to have been folded into the vadose zone transport field transport studies. Although the field study provides a valuable opportunity to test advanced characterization techniques, the magnitude of the S&T need would seem to warrant dedicated laboratory, theoretical, and field-based efforts beyond the immediate scope of the vadose zone transport field transport studies.

DISCUSSION AND RECOMMENDATIONS

In general, the research activities planned under the Vadose Zone Technical Element address important unresolved scientific issues relevant to subsurface remediation problems at Hanford. However, the technical merits of the individual projects are difficult to assess because appropriate details on the approaches to be used are frequently lacking. The different activities are well integrated, largely through a focus on the vadose zone field studies, but the direct importance of the individual studies to remediation decisions is unclear.

The planned vadose zone field studies are an important element of the research program because they integrate geochemical investigations, transport modeling, and advanced characterization techniques and provide data that can be used to evaluate upscaling methodologies. However, much of the new information that would be obtained through the S&T work reflects laboratory or small-scale field observations and consequently is not directly applicable to the large field scales pertinent to remediation. Moreover, the long period of time required to carry out vadose zone field experiments in dry environments such as Hanford is not considered adequately in the planning.

One of the main "owners" of S&T results from the Vadose Zone Technical Element will be the SAC, which can use these results to develop more realistic models for contaminant transport in the vadose zone. The hydraulic and transport parameters to be used in the vadose zone models in SAC will be derived in part from laboratory measurements on centimeter-scale core samples and will then be extrapolated to the hundred-meter scales relevant to field transport. The scientific basis of an upscaling algorithm to calculate "effective" parameters for a large block of heterogeneous sediments from highly variable measurements on small samples has not yet been developed and demonstrated. A basic problem is that small core samples cannot capture large-scale geometric features that often dominate contaminant transport in highly heterogeneous hydrogeologic settings.

Consequently, a sound upscaling framework is essential to provide the link between readily measured laboratory properties and field-scale behavior pertinent to remediaton problems, thereby establishing a basis for assessing the importance of new information in remediation decisions. However, the development of an upscaling approach that could bridge this scale gap is deferred until late in the project. The lack of early emphasis on an upscaling framework is a serious weakness of current plans because this framework should play a central role in the design of field experiments and also can be used to assess more directly the impact of new information in remediation decisions, thereby providing a basis for setting research priorities. **To address this weakness, the committee recommends that the upscaling work planned as part of the vadose zone transport field studies be initiated as soon as possible.**

The vadose zone transport field studies could provide critical data for scaling hydrologic parameters and elucidating three-dimensional flow in the subsurface at time scales relevant to site remediation. These studies are scientifically complex and costly, and their outcome could have important impacts on other Integration Project work, particularly the SAC, and on several core projects (Tank Farm Vadose Zone Project, 200 Area Remedial Action Project, Immobilized Low-Activity Waste Project). Consequently, it is essential that these studies be done well the first time.

The committee therefore recommends that peer review[7] be established specifically to provide continuing oversight of these field studies. This peer review should occur during all stages of the studies—that is, from initial planning and design of the experiments through analysis and interpretation of results.

---

[7] See Chapter 10 for a definition of peer review.

7
Groundwater Technical Element

The Groundwater Technical Element supports research on the saturated zone at the Hanford Site, especially at its interfaces with the vadose zone and Columbia River. The results of the work supported under this technical element will be used by the Hanford Site's "core" groundwater project, which is responsible for site-wide groundwater monitoring and remediation (see Chapter 3), as well as the System Assessment Capability (SAC; see Chapter 4).

Groundwater occurs beneath the entire Hanford Site, and at present, it provides the primary pathway for contaminant transport from the site to potential receptors in the river and surrounding environment. Many radionuclides of concern at the Hanford Site are highly mobile in groundwater and are transported with little or no retardation (e.g., tritium, technetium-99; see Figure 2.8a). Transport of other radionuclides by groundwater tends to be slower, either because they are less soluble (e.g., uranium, plutonium) or because they react strongly with minerals in the vadose zone before they reach the groundwater (e.g., cesium-137). Chemical contaminants such as carbon tetrachloride—a dense, nonaqueous phase liquid (DNAPL)—are only slightly soluble in groundwater. They tend to be partitioned between groundwater and a pure phase, and their presence in the subsurface can actually modify hydrologic properties (e.g., DNAPLs can partially fill pores, thereby changing water-filled porosity and hydraulic conductivity). As discussed in Chapter 2, DNAPL contamination is a serious problem in the 200 Area at the site (see Figure 2.7).

Rates of groundwater flow beneath the Hanford Site generally range from a few to several hundred meters per year, depending on hydraulic gradients and subsurface properties. At the faster rates, contaminants can be transported across the site in a few decades, which has in fact occurred for tritium (Figure 2.8a). Indeed, the groundwater pathway of particular concern at the Hanford Site stretches from the 200 Area on the Central Plateau, where most of the waste inventory and subsurface contamination exist today, to the Columbia River (see Figure 2.1), some 15-20 kilometers distant. As discussed in Chapter 2, chemical processing operations in the 200 Area resulted in the discharge of billions of gallons of water to ponds, cribs, and wells, which raised water table elevations (see Figure 2.6). These hydraulic mounds have generally accelerated flow rates and, in some cases, have reversed flow directions from natural conditions.

Groundwater tends to follow nearly horizontal flow paths in the sediments underlying the Hanford Site. Because of this, groundwater flow

Groundwater Technical Element

is often modeled as two dimensional, with no vertical structure. In detail, however, groundwater flow is three dimensional. Vertical components of flow may be substantial where contaminants enter the groundwater from disposal areas in the vadose zone. Vertical gradients influence the distribution and transport of contaminants (Figure 7.1) and complicate the task of monitoring contaminant movement in the subsurface. The difficulties arising from the three-dimensional nature of contaminant plumes in groundwater are reflected in the science and technology (S&T) plan reviewed in this chapter.

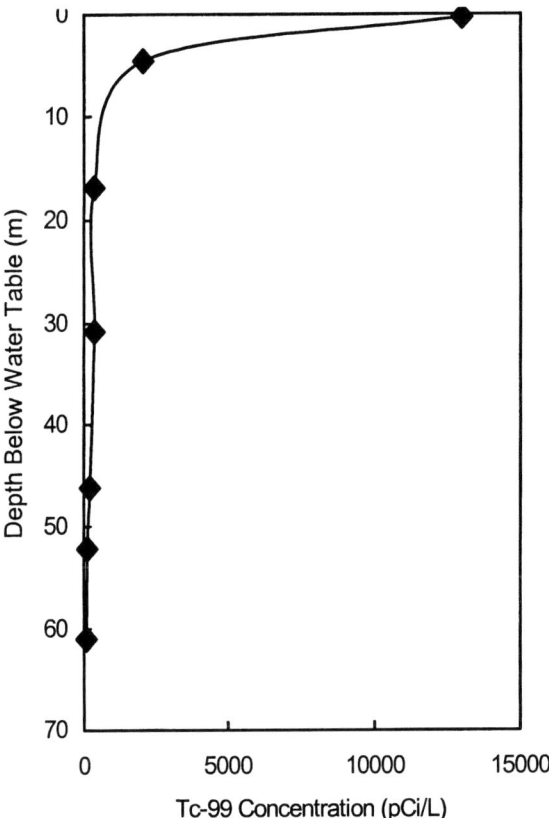

Figure 7.1 Technetium-99 concentration gradient below the water table in the 200 West Area. SOURCE: Data for well 299-W10-24, sampled October 9-16, 1998, from Hartman et al., 2000, Table 2.8-3.

Groundwater represents the saturated end member of the vadose zone, and several issues for understanding and characterizing the unsaturated zone discussed in Chapter 6 also apply here, especially with respect to hydrologic complexity and scaling relationships. However, the "upscaling problem" described in Chapter 6 for the vadose zone is of less a concern when describing physical transport mechanisms (i.e., advection, dispersion) in the saturated zone. In particular, established procedures exist for determining large-scale hydrologic properties for groundwater transport from field tests (e.g., pump tests). Moreover, hydrologic properties of the porous medium can usually be represented by single values (or tensors). Corresponding properties in the vadose zone must be represented by nonlinear functions of saturation that often exhibit hysteresis, as discussed in Chapter 6. Scaling of geochemical properties remains to be developed, but in general, groundwater transport of contaminants is easier both to measure and to model than is vadose zone transport.

SCOPE OF THE GROUNDWATER TECHNICAL ELEMENT

The main sources of information used in this assessment are the Integration Project Roadmap (DOE, 2000a) and briefings received during the committee's information-gathering meetings. The schedule and budget for S&T work under the Groundwater Technical Element are shown in Table 7.1.

The Groundwater Technical Element comprises six broad S&T activities and, within these, 20 individual projects (Table 7.1). To date, only one of these projects has been funded, as discussed in more detail below.

1. *Groundwater-vadose zone interface study.* This activity includes four projects (designated GW-1 through GW-4 in Table 4.1 in the Integration Project Roadmap [DOE, 2000a]) designed to better document the relationships between contaminant transport through the vadose zone and the consequent formation and evolution of three-dimensional contaminant plumes in groundwater. The four projects investigate three-dimensional plume structure beneath soil sites—for example, cribs and tile drains (GW-1), dilute waste tanks (GW-2), concentrated waste tanks (GW-3), and other waste sites (GW-4).

2. *Biogeochemical reactive transport.* This activity includes two projects (GW-5, GW-6) to obtain an improved understanding of the effect of redox and complexation reactions on radionuclide (mainly actinide element) transport in groundwater connected with two plutonium-bearing waste streams and one project (GW-7) to obtain an improved

understanding of multiphase reactive transport of DNAPLs, particularly carbon tetrachloride.

3. *Hydrogeological characterization study.* This activity includes five projects (GW-8 through GW-12) to evaluate the variability and scaling of subsurface hydrological parameters that control contaminant transport. One project (GW-8) will approach the problem using existing groundwater data. Two projects (GW-9 and GW-10) will conduct otherwise unspecified "multiple-scale studies" beneath clean and contaminated sites. The remaining two projects (GW-11 and GW-12) will synthesize data and construct three-dimensional visualizations of hydrogeological properties under soil and tank sites.

4. *Regional plume geometry.* The activity includes one project (GW-13) to develop a three-dimensional image of contaminant plumes along a transect extending from the 200 West Area to the Columbia River.

5. *Multiscale three dimensional model development.* This activity includes three projects (GW-14 through GW-16) to develop approaches for implementing three-dimensional transport models that can be run at multiple scales and three projects to develop methods for incorporating heterogeneity and uncertainty in these models for groundwater under a boiling waste tank (GW-17), a specific retention basin (GW-18), and a dilute waste tank (GW-19).

6. *Groundwater discharge study.* This activity includes one project (GW-20) to quantify the three-dimensional plume dynamics at a site along the Columbia River.

As shown in Table 7.1, work on these projects is planned to run from fiscal year 2000 through fiscal year 2004. The total planned funding for this technical element is about $16.3 million. Note, however that these funding levels, which were identified in the roadmap, have been revised by the Department of Energy (DOE) (see Table 3.1).

EVALUATION OF WORK PLANNED UNDER THE GROUNDWATER TECHNICAL ELEMENT

As of early 2001, only the "groundwater discharge" study (GW-20; see Table 7.1), which is concerned with groundwater in the 100 Area, had been initiated; it was supported by funding from Hanford's "core" groundwater project (Chapter 3). Consequently, there was little scientific or technical output in the form of peer-reviewed reports or papers available for the committee's evaluation. Accordingly, the committee offers only general comments about the work planned under this technical element, again focused on the five evaluation questions against which the other S&T elements are compared. The lack of specificity in the

TABLE 7.1 Summary of S&T Activities and Planned S&T Projects Under the Groundwater Technical Element

S&T Activity	S&T Projects Planned	Project Objectives	Project Duration (fiscal years)	Hanford Funding[a] (thousand dollars)	EMSP Funding (thousand dollars)
Groundwater-vadose zone interface study	4	Obtain an improved understanding of the relationships between contaminant transport through the vadose zone and plume formation in groundwater	2001-2003	2,100	0
Biogeochemical reactive transport	3	Obtain an improved understanding of redox conditions, the role of complexants in transport, and the location and characteristics of DNAPL contamination	2001-2004	7,800	0
Hydrogeological characterization study	5	Develop an improved understanding and characterization of subsurface heterogeneity on contaminant transport	2001-2003	3,000	0
Regional plume geometry	1	Obtain an improved understanding of the three-dimensional geometry of contaminant plumes in groundwater	2001	2,100	0
Multiscale three-dimensional model development	6	Obtain an improved understanding of heterogeneity and uncertainty that can be incorporated into multiscale models	2001-2003	1,300	0
Groundwater discharge study	1	Obtain an improved understanding of contaminant release locations and fluxes to the Columbia River	In progress	0^b	0

NOTE: EMSP = Environmental Management Science Program
[a] The Integration Project intends to seek funding from national S&T programs (e.g., DOE Headquarters) for some of this work.
[b] The River Monitoring Project (see Chapter 3) is providing funding and leadership for this work.
SOURCE: DOE, 2000a, Figure 4-1, Table 5-1.

Groundwater Technical Element 105

information available in the groundwater portion of the Integration Project Roadmap precludes a more detailed assessment.

Can the objectives of the planned work be achieved?

The committee found it difficult to provide a definitive answer to this question because of the lack of technical detail on the planned projects in the Integration Project Roadmap (DOE, 2000a). The S&T objectives may be achievable if the planned work is funded at adequate levels and tied to site decisions. However, some fundamental issues must be resolved to bring these tasks to completion.

In particular, the Integration Project may have unrealistic expectations about the time that will be needed to complete some of these studies. For example, task GW-11 (Synthesis and Visualization of Hydrogeology—Soil Site) has a 10-month time line. Task GW-12, which has the same general objective for a high-level waste tank site, has a 24-month time line. The expected outcomes of GW-11 and GW-12 include providing estimates of small-scale hydrogeological property variability and spatial correlation in a form amenable for use in numerical models, investigating the scale dependence of hydraulic measurements, and investigating important scales of physical and hydrogeological heterogeneity characterization.

Issues of scaling that are raised within the scope of these projects are an active focus of research efforts in many scientific disciplines (see Sidebar 6.1 and an expanded discussion of scaling in Appendix C). It is probably more realistic to anticipate that significant progress on scaling issues will be measured on a time scale of 5 to 10 years, rather than the 1- to 2-year time frames allowed for these projects, even if funded at the requested levels. The Integration Project should consider the implications of slower-than-planned progress on these projects for other work at the site (e.g., the SAC) and should adjust the schedules accordingly, if appropriate.

Does the planned work represent new science?

Again, the lack of detailed information makes it impossible for the committee to identify specific areas of new science. However, opportunities appear to exist to develop new understanding and better quantification of issues such as the three-dimensional nature of contaminant plumes and hydrogeological characterization, both of which are identified in Table 7.1. The research planned on each of these topics is generalizable beyond Hanford. Underlying questions and anticipated outcomes apply in a broad sense to many contaminated sites where remediation and stewardship are planned or under way. It is critical, however, that this research be conducted at Hanford, in light of the mix of contaminants that have been released to the environment and with respect to transport to and interactions with the Columbia River.

Can the planned work have an impact on cleanup decisions at the Hanford Site?

Better characterization of groundwater pathways and contaminant fate and transport in the saturated zone has obvious relevance to issues of site remediation and long-term stewardship, especially in the 200 Area. It is clear, however, that the understanding of groundwater flow and transport is more mature than that for vadose zone flow and transport. Consequently, uncertainties in groundwater models at the Hanford Site are small relative to uncertainties in vadose zone and river models. Therefore, S&T directed at refining the understanding of groundwater transport may not be a good investment relative to S&T efforts that are needed to improve the understanding of vadose zone and river transport (see Chapters 6 and 8).

Does the planned work address the important issues?

The broad tasks outlined in the Groundwater Technical Element address the core issues that have to be resolved with respect to contaminant fate and transport in groundwater at the Hanford Site. In the course of these studies, sophisticated computational tools may be developed that can aid in making sound site management decisions. Valuable basic data on the hydrogeology of the saturated zone and contaminant distribution in the groundwater system also may be obtained. These data may be important for achieving progress in a number of other site projects, such as the development of long-term monitoring plans for the groundwater system at Hanford.

Are there other concerns, comments, or suggestions that should be considered by the Integration Project in executing the planned work?

The committee has two concerns. First, as noted previously, detailed project descriptions do not appear to exist in many cases, and written descriptions of the projects in the Integration Project Roadmap (DOE, 2000a) were too brief to determine how likely it is that the projects will meet their objectives. Second, although the projects may provide valuable contributions to science, it is not clear whether the S&T results are needed for site decision making. Essentially all of the projects are assigned to the priority ranking "Critical to the success of the Accelerated Cleanup: Path to Closure" project in the Integration Project Roadmap

Groundwater Technical Element

(DOE, 2000a). The discussion in Appendix B of DOE (2000a) of the consequences of not filling a particular research need, although valid, provides little substantive information to guide a prioritization effort. **Therefore, the committee recommends that a more selective system of prioritization be developed for these projects and that each project be referenced to this prioritization system before subsequent funding cycles begin.** A more detailed discussion of prioritization is provided in Chapter 10.

DISCUSSION

The Integration Project has clearly assigned a lower priority to the Groundwater Technical Element than to the Vadose Zone Technical Element, as shown by the planned funding levels and schedules in Tables 3.1, 6.1, and 7.1. According to the Integration Project Roadmap (DOE, 2000a), there were plans to start 14 of the 20 groundwater activities by February 2001. As of March 2001, only one groundwater activity (GW-20) was under way.

Although the documentation of detailed research plans is sparse, the planned S&T activities in the Groundwater Technical Element appear to identify a set of projects and investigations that can add confidence to the assessment of contaminant migration in groundwater at Hanford. Because groundwater modeling has progressed to a greater degree than many other S&T issues discussed in this report, the committee agrees with the Integration Project's decision to assign a lower priority to the Groundwater Technical Element relative to the other technical elements. The committee notes, however, that the basis for this decision does not appear to be documented and was therefore not reviewable.

Among the activities included with the Groundwater Technical Element, assignment of the highest priority to the groundwater-river interface study (GW-20) is clearly driven by the intensive restoration efforts under way along the Columbia River (see Chapter 2). This research activity is likely to have a more immediate return in better managing current cleanup activities along the river corridor than the other projects planned under this technical element.

8
Columbia River Technical Element

This chapter provides a review and assessment of Integration Project science and technology (S&T) under the Columbia River Technical Element. This technical element supports studies to improve understanding of the river environment, particularly as it relates to contaminant inputs, transport, and impacts on biological systems. The Columbia River is likely to be the main pathway for contaminant transfer to humans because local populations rely on the river to various degrees for recreation, irrigation and drinking water, and other sustenance. Thus, to the extent that contaminants are present or are likely to be introduced into the Columbia River, understanding their fate and transport in the river system is important for protecting present and future human populations.

COLUMBIA RIVER HISTORY[1]

The Columbia River drains an area of 259,000 square miles (39,000 square miles in Canada). It is the fifth largest river in the United States in terms of area and the third largest in terms of discharge. There are more than 400 dams in the watershed that provide more than 21 million kilowatts of electrical power generating capacity—including 11 dams on the Columbia River, 7 of which are located upstream from the Hanford Site and four downstream (Figure 8.1). Damming of the Columbia was initiated in 1938 with the completion of the Bonneville Dam and ended in 1973 with the completion of the Mica Dam in Canada.

The Hanford Reach extends from the Priest Rapids Dam upstream of the Hanford Site to the head of Lake Wallula, which was created by the McNary Dam (Figure 8.2). It is the last free-flowing stretch of the Columbia River, although discharge through the reach has been altered by upstream controls. The historical variation in river stage through the reach of up to 27 feet has been reduced to 9-10 feet by dam construction.

---

[1]The information in this section was taken from several reports and papers, including Rickard and Watson, 1985; Minshall, 1988; Rickard and Gray, 1995; Zorpette, 1996; Williams et al., 1998; the Center for Columbia River History [http://www.ccrh.org/river/history.htm]; and the Pacific Northwest National Laboratory [http://www.pnl.gov/env/surface-water_surveillance.htm].

Columbia River Technical Element

From a historical perspective, the environmental characteristics of Hanford Reach can be divided into three periods, all of which are related to differing land use activities. The first period is pre-dam construction (prior to 1930), the second occurs during Hanford Site operations (1944-1980), and the third includes post-Hanford operations and river restoration (1980-present).

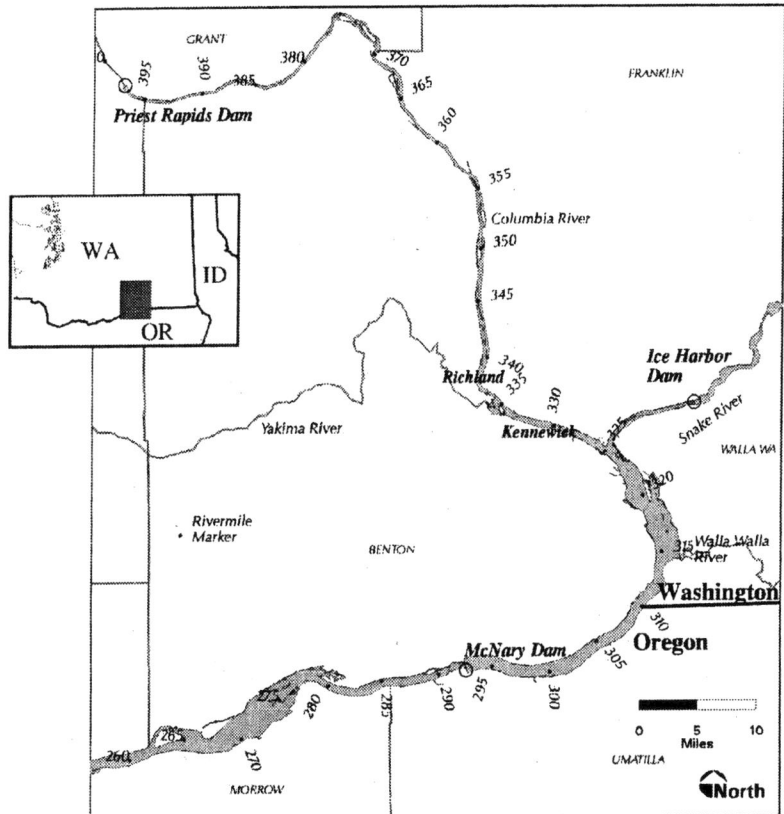

Figure 8.1 Plan view of the Columbia River in the vicinity of the Hanford Site. Numbers denote miles above the river mouth. SOURCE: BHI, 1999, Figure E-5.

During the pre-dam construction period, the area around the Hanford Site was occupied for at least 10,000 years by several Indian groups. The first Euro-American exploration in the area was by Lewis and Clark in 1805. Indian groups ceded their lands to the government at the Treaty Council of 1855, leading to the expansion of Euro-American settlements.

By 1860, a ferry was operating across the Columbia River at White Bluffs, one of the first permanent settlements in the Hanford area on the east bank of the Columbia River. General population increases with the influx of gold miners at about that same time encouraged ranching development across the Columbia Plateau. Steamboats were also operating on the Columbia River up to White Bluffs.

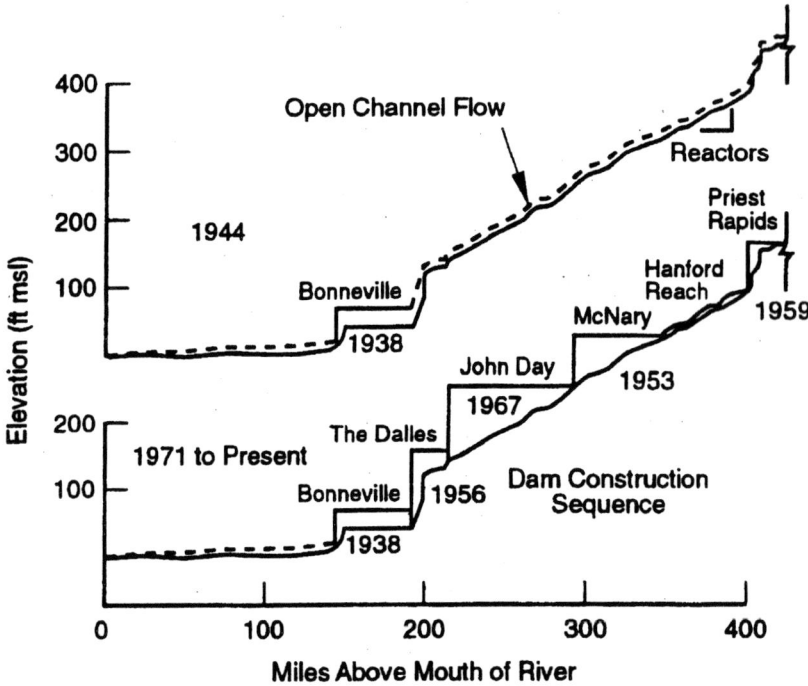

Figure 8.2 *Top*: Profile of the Columbia River channel bed in 1944. *Bottom*: Profile of the river channel at present. Dates on the figure indicate dam completion. NOTE: msl = mean sea level.
SOURCE: BHI, 1999, Figure E-4.

Arid conditions in the region prevented the spread of dryland farming, and agriculture was limited until irrigation canals were constructed in the late 1880s and further expanded in the early 1900s. Railroad construction led to the founding of Pasco and Kennewick in the early 1880s, but difficult weather and soil conditions frustrated the development of the area through the early 1930s. Irrigation projects bringing water to the north and east of the Hanford Site were completed in the 1950s and continue to support agricultural activity to the present.

Although few water quality data are available prior to 1930, the lack of development and limited use of the river suggest that water quality was good. The natural flow variability limited riparian vegetation and contributed to varied habitat conditions supporting more than 44 species of fish in the Hanford Reach.

With initiation of construction on the Hanford Site, small settlements were relocated and site worker populations were as high as 50,000 in the 1940s. Water quality records for discharges to the Columbia River become available with the advent of operations at the site. Beginning in 1944, operating reactors released heated water, radionuclides, and corrosion-inhibiting chemicals directly to the river (see Chapter 2). The work force of 50,000 generated domestic waste that was also discharged to the ground and river.

Between 1951 and 1965, the operating reactors released a maximum of 24,000 megawatts of heat and 10,000-12,000 curies of radionuclides to the river each day. It is estimated that 110 million curies of radiation were released to the river between 1944 and 1971 (Rickard and Watson, 1985; see Chapter 2). Most of the released radionuclides had short half-lives, and releases included activation products associated with natural elements present in the cooling waters. These discharges introduced short-lived radionuclides such as phosphorus-32 and zinc-65 into river biota. Longer-lived radionuclides, now buried in river sediment, include cobalt-60, strontium-90, cesium-137, uranium-238, and plutonium-238, 239, and 240. In addition to releases from the Hanford Site, irrigation return water from agricultural activity to the east of the site entered the Columbia River (as both surface and groundwater) near Ringold and Richland.

Between 1964 and 1971, the single-pass plutonium production reactors were phased out (see Table 2.1) and chemical and radionuclide releases to the Columbia River essentially ceased. The N-Reactor continued to release heat, but little radioactivity. It was shut down in 1987.

In the restoration phase that followed the shutdown of all production reactors, environmental surveillance work has found that radionuclides in biota in the Hanford Reach are the same as in biota from upstream reference reaches of the Columbia River where no radiation was released. The major environmental concern for river ecosystems is the regulation of flow from upstream dams. The lack of high-stage variability has altered riparian vegetation, and the presence of artificial water-stage controls has led to destruction of spawning redds (Figure 8.3) and stranding of fish, because stage variations now occur over hours rather than days to weeks as in the past (Williams et al., 1998). The major continuing concerns about water quality in the Hanford Reach include the addition of agricultural chemicals, discharge of effluents from upstream industrial development, and the continuing contribution of low levels of contaminants from the Hanford Site through groundwater discharges into the river.

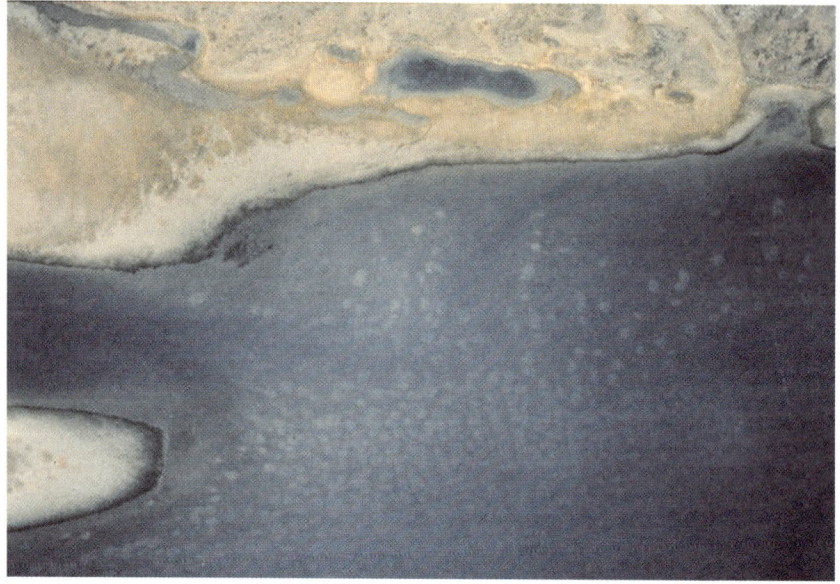

Figure 8.3 Photo of spawning redds in the Hanford reach in a slough immediately downstream of the 100-F Reactor area. Downstream is to the left. SOURCE: Zack Carter, Boeing Computer Services.

SCOPE OF THE COLUMBIA RIVER TECHNICAL ELEMENT

The Columbia River Technical Element comprises five S&T activities with 21 individual projects (Table 8.1):

1. *Detailed conceptual model.* This activity includes three projects (CR-1 to CR-3) intended to develop a conceptual model that accounts for contaminant fate and transport in the river system.
2. *Information management.* This activity includes four projects (CR-4 to CR-7) intended to develop a system to gather, screen, and manage data for the river assessment and also to populate this system with available data and information from both the Department of Energy (DOE) and external sources.
3. *Characterization.* This activity includes four projects (CR-8 to CR-11) intended to identify habitats, species abundance, and distributions and to determine biological transfer functions[2] for contaminants-species combinations of interest.
4. *Groundwater-river interface study.* This activity comprises six projects (CR-12 to CR-17) intended to develop and test conceptual and numerical models for groundwater and contaminant discharges to the Columbia River.
5. *Fate and transport.* This activity comprises four projects (CR-18 to CR-21) intended to develop and test conceptual and numerical models for fate and transport of contaminants in the river system.

The schedule and budget for the Columbia River Technical Element are given in Table 8.1. Projects are planned to run from fiscal year 1999 through fiscal year 2004, and work was under way in four of the five activities during the committee's review. The total planned funding for this technical element is $8.17 million, but the Integration Project plans to obtain at least half of this funding from Hanford's core programs, and some of the future funding may be provided by external sources such as DOE Headquarters.

---

[2] A biological transfer function is a measure of the movement of a contaminant between food-chain levels in an ecosystem—for example, the transfer between microscopic algae growing on rock surfaces in the river and the aquatic insects that graze on those algae.

TABLE 8.1 Summary of S&T Activities and Planned S&T Projects Under the Columbia River Technical Element

S&T Activity	S&T Projects Planned	Project Objectives	Project Duration (fiscal years)	Hanford Funding (thousand dollars)	EMSP Funding (thousand dollars)
Detailed conceptual model	3	Develop a detailed conceptual model of the river system that includes critical components and processes and identifies important links	1999-2002	225[a, b]	0
Information management	4	Develop an information management system to gather, screen, and manage data and information for river assessment and populate this system with available data and information	2001-2002	325[b]	0
Characterization	4	Identify habitats, species abundance, and distributions and determine biological transfer functions for contaminant-species combinations of interest	2000-2004	3,520[a, c]	0
Groundwater-river Interface study	6	Obtain an improved understanding of, and develop and test conceptual and numerical models for, contaminant discharge from groundwater to the Columbia River	1999-2004	2,000[a, d]	0
Fate and transport	4	Develop and test conceptual and numerical models for contaminant fate and transport in the river system	1999-2004	2,100[a]	0

NOTE. EM3P – Environmental Management Science Program
[a]Additional funding for this work is being provided through the System Assessment Capability (see Chapter 4).
[b]The funding shown in the table will be provided by the Characterization of Systems Project (see Chapter 3).
[c]The funding shown in the table will be provided by the River Monitoring Project (see Chapter 3).
[d]The Integration Project intends to seek funding from national S&T programs (e.g., from DOE Headquarters) for some of this work.
SOURCE: DOE, 2000a, Figure 4-1, Table 5-1.

EVALUATION OF WORK PLANNED UNDER THE COLUMBIA RIVER TECHNICAL ELEMENT

The committee's assessment of this technical element is based primarily on a review of the Integration Project Roadmap (DOE, 2000a) and planning documents (DOE, 1999e; DOE, 2000g) provided by the Integration Project and secondarily on information received at committee meetings. The latter includes a demonstration of a numerical model of the groundwater-river interface, discussions with several Integration Project investigators, and (for one committee member) a tour of the Hanford Reach.

The committee provides assessments below of each of the S&T activities shown in Table 8.1. Three of the assessments are structured using the five evaluation questions that were first introduced in Chapter 6. There was not enough information available to use these questions to structure the remaining two assessments, so the committee instead provides brief commentaries.

Conceptual Model Development

The S&T projects under this activity are focused on the development of a river conceptual model. This model will provide a quantitative description of the processes that occur in various components of the river system, including the riparian zone and associated biota along the river, aquatic biota, groundwater-river interface, river bottom and sediments, river water column, and users of river resources. The model for this element is termed "conceptual" because it is focused on identifying important processes and links among the various model components listed above, and it will connect processes in the Hanford Reach with important river controls both upstream and downstream of the site. Once the important processes are identified, a numerical model that can simulate these processes will be developed.

The conceptual and numerical models are being developed in three modules: (1) the zone of groundwater-river interaction; (2) hydrodynamic, sediment, and contaminant transport in the river; and (3) biological transport. The Integration Project plans to use the model results for site and downstream risk and impact assessments. The model is designed to estimate concentrations for four classes of radionuclides and two chemicals. The System Assessment Capability (SAC) will be the primary user of this model (DOE, 2000g).

Can the objectives of the planned work be achieved?

The Columbia River conceptual model as described in the SAC documents (e.g., BHI, 1999) is intended to account for important contaminant transport processes in the Columbia River. The inclusion in this model of regional-scale river processes is, in the committee's view, essential to obtain useful impact assessments, because conditions in the river at the Hanford Site are controlled by upstream dam operations as well as water quality management activities in the watershed.

Based on the committee's understanding, this conceptual model can probably support SAC Rev. 0 needs (see Chapter 4). The conceptual model may have limited resolution, however, because it models the river as a series of segments, each characterized by a set of average model parameters. The sizes of these segments were unspecified in the materials reviewed by the committee. The general structure of the conceptual model appears to the committee to be robust and may support expansion as needed to meet other, possibly as-yet-undefined, objectives in the future.

Does the planned work represent new science?

The approach being taken is best described as the application of current science and modeling techniques, rather than new science. Many components of the model have already been developed and/or applied to the Columbia River. On the other hand, there are not, to the committee's knowledge, many models that integrate the complex linkages among the major components (i.e., groundwater, river hydrodynamics, biological receptors) of large river systems. In the committee's opinion, this conceptual modeling effort at Hanford, if successful, is likely to contribute to the development of capabilities that can be applied to other large rivers.

Can the planned work have an impact on cleanup decisions at the Hanford Site?

Although the river is a critical and visible element of the cleanup program, the Hanford Reach has already shown marked recovery from past contaminant discharges (see Chapter 2 and the discussion elsewhere in the chapter) because of its high capacity for dilution and transport. The committee expects that this conceptual model will confirm the Columbia River's capacity for self-maintenance and may support the prioritization of resources to those areas of the Hanford Reach subject to greatest contaminant loadings—for example, spatially restricted zones of contaminated groundwater discharge to the river bed. However, the present level of resolution of the model will likely be inadequate to support impact assessments or decision making in these spatially restricted zones.

Does the planned work address the important issues?

The planned work is designed to develop a conceptual model that will support site decision making. The model may assist in impact analysis and contribute to more effective management of contaminants at the Hanford Site. As noted above however, the level of resolution in the model will limit its utility in addressing site-specific contaminant issues in the Columbia River.

Are there other concerns, comments, or suggestions that should be considered by the Integration Project in executing the planned work?

The committee has two general concerns about this S&T activity: the first is model resolution or scale, and the second is model validation. The concern related to model resolution involves the range of scales over which the model must operate. This issue of scale is addressed, in part, in discussions provided elsewhere in this report (see, for example, Chapter 6 and Appendix C). As noted previously, the plan to model the river as a series of segments, the lengths of which were undocumented in the materials reviewed by the committee, probably will preclude the use of this model to assess impacts from specific contaminant discharge zones in the bed of the Columbia River.

The concern about model validation is also related to scale. Temporal and spatial model scales may range over 13 orders of magnitude. The selection of a single or even a limited number of scales for model analysis will affect model validation efforts. For example, the scale of the data to be used in model calibration and validation may not exactly match the scale of resolution of the model itself, particularly when historic data, which were not collected specifically to meet model needs, are used. Further, the inherent variability of large river ecosystems will complicate normal model validation efforts. Although the concerns about scale and validation raised here are relevant for any model development and application effort, the time scales associated with contaminant effect and the spatial scales associated with contaminant movement and effect suggest that these issues will be of particular importance to Columbia River modeling.

Information Management

Environmental data have been collected at the Hanford Site since it became operational in 1944. Additional monitoring programs have been put into place in succeeding decades in response to regulatory requirements, and these continue to generate large amounts of data. Recognizing the extent and complexity of existing data resources, the objective of the four S&T projects (CR-4 to CR-7) under this activity is to develop an information management system to "screen, manage, and disperse" data and information from both inside and outside Hanford. The information management system will be developed by September 2001 and will be updated regularly as additional site data are collected.

The committee recognizes the need for efficient and effective data management at the Hanford Site. There may be S&T needs associated with the development and application of new information technologies, but these were not made clear in the documentation reviewed by the committee. Therefore, given the apparent lack of an S&T context for this work, information management development would seem to be better handled in programs other than the Integration Project S&T program. With completion of the planned information management system later this year, additional S&T work does not appear to be needed on this issue.

Characterization

The four characterization projects (CR-8 to CR-11) under this activity involve fate and transport model parameterization and environmental data collection to support fate and transport analysis for future SAC revisions (see Chapter 4). The projects under this activity will elucidate the transfer of contaminants through organisms and the identification of critical habitats. Because SAC revisions will require increasingly complex and detailed data, the characterization projects must meet future as well as present SAC data needs.

Although details of future SAC revisions are not available at this time, the committee expects that model revisions will involve reductions in model scales (both spatial and temporal) to obtain increases in model resolution. The improved resolution may allow the models to be used to assess impacts at more spatially restricted river scales than is possible with the current generation of models.

In general, the planned projects appear to the committee, at least on the surface, to be designed to respond to expected data needs. However, the available project descriptions are inadequate to allow detailed evaluations of individual projects. It is critical that these projects be integrated effectively with those from the other technical elements to ensure that the data collected meet the "characterization" needs of those projects or provide sufficient information to support model calibration and validation.

In the committee's view, an important S&T need under the characterization activity is to define and quantify impact "thresholds," that is, points at which small changes in environmental conditions can result in major changes to species habitat, abundance, or health. Similarly, the committee believes that the characterization activity should provide the Integration Project with techniques or characterization protocols that could support the development of new tools for impact assessment. The committee did not see this outcome explicitly identified, but documentation did suggest that improved impact assessment would be an outcome of this activity.

Considering the scope of future management needs in the Hanford Reach, and the lack of full understanding of ecosystem structure and function in large river systems, the committee believes that new monitoring tools and techniques may be required to obtain the needed characterization data. The development of new large river and ecosystem monitoring tools—specifically to provide information about organism distribution and ecology in the river channel and the interaction between the river and the riparian zone—is viewed by the committee as an important S&T gap in the current program.

Although there are numerous biological monitoring tools available (Schaeffer and Herricks, 1993), these tools have limited application to ecosystem dynamics (Schaeffer et al., 1988). This is particularly true for large river systems, the theoretical ecological foundations for which are based on research in small streams. Similarly, the interactions between rivers and riparian/floodplain ecosystems is an important element of river ecology (Ward, 1989) that is poorly understood for large rivers. It should be recognized that ecology is entering a new era of analysis supported by advancing technology (Thompson et al., 2001), which will allow researchers to address how biological and physical processes interact over multiple spatial and temporal scales. The committee believes that the S&T program should be oriented to advancing the capability for impact assessment in the Columbia River using the most modern monitoring and analysis tools available.

The committee therefore recommends that the S&T program support the development of advanced biological monitoring approaches for the Columbia River. Additional details on this recommendation are provided in the last section of this chapter.

Groundwater-River Interface

The groundwater/river interface activity includes six projects (CR-12 to CR-17) that are designed to model the interactions between groundwater and the Columbia River. The projects are focused on developing models and filling data needs by conducting field and complementary laboratory experiments. The planned investigations include the elucidation of dynamics of flow direction; contaminant attenuation, decay, and transformation; biological processes; transport rates; and preferential pathways associated with contaminant discharge from aquifers to the Columbia River. The models developed in these projects will provide the primary support for contaminant fate and transport analysis and eventual prediction of impacts. Planned activities include numerical model development and simulations to develop impact predictions. Based on DOE (2000g), it appears that the first model development will be completed in fiscal year 2002.

Can the objectives of the planned work be achieved?

The groundwater-river interface activity is designed to support fate and effects analyses for the Columbia River. Although the documentation suggests a 2002 start date for this activity, as noted above, a model that integrates groundwater flow with river stage changes was demonstrated to the committee during one of its meetings, which suggests that progress already is being made on this activity.

Some of the objectives of the planned work in project CR-13 depend on the completion of several field and laboratory investigations that have not yet begun. Planning documents reviewed by the committee suggest that these investigations are planned to be completed in approximately six months. This schedule is overly optimistic given the complexity of the planned work. Further, the projects that are supposed to compare predictive and observational data against impact criteria (CR-15 to CR-17) do not have clear plans or objectives. Without clearly defined objectives, it is not possible to evaluate project merit or to predict success.

Does the planned work represent new science?

The numerical model to be developed for the groundwater-river interface is unusual because of the river size and the rapid flow alterations due to upstream dam releases, which produces rapid changes in river stage, changes in hydrostatic regimes in the river banks, and corresponding changes in groundwater movement. This model will be able to build on the groundwater and vadose-zone modeling under way at the Hanford Site in that these models will provide good estimates of the rate of delivery and expected concentration of contaminants to the river bank area. This work has the potential to produce new insights, and the modeling approaches developed are potentially applicable to other large rivers.

Can the planned work have an impact on cleanup decisions at the Hanford Site?

This model may allow the Integration Project to describe more effectively the interactions between groundwater and the Columbia River, which can contribute to understanding contaminant impacts on the river system. This model could lead to improvements in prioritization of cleanup activities based on future hazards to the Columbia River, if the model resolution issues raised previously are addressed.

Does the planned work address the important issues?

Given the fact that groundwater discharges along the Hanford Reach are on the order of 10^1-10^2 cubic feet per second and the average river discharge through the reach is on the order of 10^5 cubic feet per second, it is unlikely that groundwater discharges will substantially affect the Columbia River at the scale of the Hanford Reach—this is true even for low-flow conditions, because river discharge is maintained by releases from upstream dams.

This said, however, it is likely that groundwater contamination will enter the Columbia River in the future in spatially limited areas, as is observed today, creating locally high concentrations in or near habitats of important organisms (e.g., salmon) in the river ecosystem. Contaminant concentrations that affect river organisms at spatially restricted scales may not produce significant impacts on the Columbia River ecosystem as a whole. Nevertheless, it will be important to identify groundwater inflow locations to assess whether any effects on critical species or habitats should to be minimized. The proposed modeling activity could contribute to the identification of both critical habitats and species that may be affected by groundwater-related contaminants entering the river in the future.

Fate and Transport

The fate and transport activity comprises four projects (CR-18 to CR-21) that involve the development of a quantitative model for hydrologic, sediment, and contaminant transport. The activity will also include laboratory-based testing that will be concluded in 2002. Descriptions of these projects provided in DOE (2000g) suggest a focus on tracking contaminant transport for only a limited list of contaminants (described as nutrient and nonnutrient metals) in a limited selection of species (one fish and one aquatic invertebrate species). The contaminants and species to be studied will be identified during the projects.

Can the objectives of the planned work be achieved?

Hydrologic models have been developed already by several organizations, and these models are being modified for application to the Columbia River. Therefore, the committee believes it is likely that the hydrologic modeling objectives of these projects will be achieved.

The fate and transport activity also includes modeling of sediment and biological transport, as well as general contaminant transport in the Columbia River. The entire set of models will be evaluated against measurements and monitoring data from the Columbia River. The committee believes that it is likely that models that capture general process characteristics can be developed to meet the objectives of this activity. The committee believes it unlikely, however, that models incorporating even a moderate level of complexity can be developed in the time frame proposed.

Does the planned work address the important issues?

Contaminant transport through the Columbia River system is an important issue, which must be addressed through both modeling and data collection. An important use of models is to identify the critical data needed for site decisions. The committee believes that this modeling exercise, if executed well, could benefit existing river monitoring programs by contributing to a better review of data utility.

This modeling exercise could also contribute to a better understanding of the possible implications of small-scale, or localized, contamination for broader ecosystem conditions in the river, if the models have the appropriate spatial and temporal scales. For example, if this modeling identifies transport pathways for contaminants or the propagation of effects from localized contaminant zones in the river, then the planned work may provide critically needed information to guide future research and data collection. The committee believes that these types of simulation models are best used to guide research and are much less useful for prediction of impacts.

Are there other concerns, comments, or suggestions that should be considered by the Integration Project in executing the planned work?

The major concern associated with the fate and transport of contaminants from the Hanford Site is the potential for contaminants to produce harm for thousands of years. In predicting fate and transport, the models must account for long-lived contaminants. Further, the models must consider the effect of extremely rare events, such as catastrophic floods, on fate and transport. Comments on these issues are provided in Chapter 4; see also Chapter 9.

SUMMARY AND DISCUSSION

The committee believes that the S&T activities planned for the Columbia River Technical Element have the potential to contribute to a better understanding of the Columbia River, the interactions between groundwater flows from the Hanford Site and the river, and the potential effects of contamination from the Hanford Site on the river ecosystem. However, in many cases the documentation was insufficient to allow the committee to judge whether this potential is likely to be realized.

The committee believes that the activities in the Columbia River Technical Element should be viewed in the broader context of the Columbia River watershed and the Hanford Reach. As noted at the beginning of this chapter, the Hanford Reach (Figure 8.2) is the last free-flowing stretch of the Columbia River. As the last remnant of a once-vast and important resource, the Hanford Reach certainly deserves attention from the Hanford Site's cleanup program—although it may well be that the impacts of site contaminants on the reach are insignificant, and therefore the river may not be the most important factor in site cleanup decisions.

The Columbia River has undergone many changes over the last 100 years. It has been dammed, its flow has been altered, and its major fishery has been compromised. At the same time, the Columbia River has been subject to high thermal, radionuclide, and other contaminant loads from the Hanford Site, with little apparent residual effects (e.g., PNNL, 1999; see also the discussion of Columbia River history elsewhere in this chapter). The Hanford Reach has been flushed by large volumes of water in the decades following the cessation of plutonium production operations at the site, and the Columbia River ecosystem has recovered through natural processes. Today, the Hanford Reach is one of the most valued and ecologically important stretches of the Columbia River.

In light of the importance of the Columbia River, the committee recognizes the need for an integrated site and river model but cautions against overdependence on modeling, where scale and validation issues can compromise model utility. The committee sees a clear need for information management, but recognizes that information management will operate in support of other cleanup activities, rather than leading them. Similarly, characterization activities, the groundwater-river interface analysis, and fate and transport modeling are viewed by the committee as supporting, rather than leading, activities of the Integration Project.

The committee has identified additional S&T needs in the Columbia River Technical Element. In particular, the committee finds a need for advanced modeling that will integrate results over the scales of analysis and the biological effects present in the Hanford Reach-Columbia River watershed. The committee supports better characterization of the environmental setting and better fate and transport information for contaminants and ecosystems. The committee recognizes, however, that limited resources may well be allocated to characterization of, or modeling contaminant transport from, areas of contaminant concentration such as the 200 Area. In other words, the committee would not place a priority in the S&T program for work in the Columbia River Technical Element given S&T needs of other technical elements.

As noted elsewhere in this chapter, **the committee recommends that the S&T program support the development of advanced biological monitoring approaches for the Columbia River.** The committee believes that there is a critical need to develop new technologies for ecosystem assessment that will complement and expand the current radionuclide and chemical monitoring capabilities. There is an extensive literature on biological monitoring that provides a clear indication of the value of using biota as sensitive indicators, for example, of water quality (Rosenberg and Resh, 1993; Kerans and Karr, 1994). Advanced biological monitoring procedures, which include a wide range of genetic, biochemical, organism, community, and ecosystem metrics, some measured in real time, represent the state of the art in assessing ecosystem quality and condition.

Finally, the committee suggests that protection of the Columbia River may benefit from a monitoring program directed to the zones of effect in the groundwater-river interface. Biological monitoring of these areas will give more power to detect contaminant impacts on the biota relative to reach-scale assessments. Early determination of impact, or a finding of no impact, in the Columbia River would provide critical information for cleanup decisions at the Hanford Site.

9
Monitoring, Remediation, and Risk Technical Elements

As noted in Chapter 3, detailed science and technology (S&T) plans for the Remediation, Monitoring, and Risk Technical Elements were being developed during the committee's information gathering for this report. S&T plans for the Remediation and Monitoring Technical Elements will not be issued until fiscal years 2002 and 2003,[1] respectively, and the S&T plan for the Risk Technical Element exists only in draft form (DOE, 2000d).[2] Consequently, the committee was unable to obtain detailed descriptions of the projects under these technical elements, and it is therefore able to provide only a general overview and assessment of the planned work.

The main sources of information used in this assessment are the Integration Project Roadmap (DOE, 2000a) for the Remediation and Monitoring Technical Elements and that roadmap and subsequently issued risk S&T plan (DOE, 2000d) for the Risk Technical Element. A summary of the S&T activities and projected budgets are given in Table 9.1.

MONITORING AND REMEDIATION TECHNICAL ELEMENTS

The Monitoring Technical Element and the Remediation Technical Element have only one planned S&T activity each, and they are scheduled for completion in 2003. The S&T activity for the Remediation Technical Element, *identification, development, and deployment of improved groundwater remediation strategies,* includes two projects to develop an improved technical basis for remediation of contaminant plumes at the Hanford Site. In the first project (Rem-1[3]), the distribution of carbon tetrachloride plumes in the 200 West Area (see Chapter 2) will be investigated, and the results will be used to assist in the development of a strategy for corrective actions. The planned work is to include geophysical, geochemical, and modeling studies to assist with the selection and deployment of remediation technologies.

---

[1] These plans were originally scheduled to be issued in fiscal years 2001 and 2002, respectively, but were delayed because of funding cutbacks.
[2] The committee received a copy of this draft in October 2000.
[3] As noted in previous chapters, the projects under each of the six activities are given these identification numbers in DOE (2000a, Table 4-1).

TABLE 9.1 Summary of S&T Activities and Planned S&T Projects Under the Monitoring, Remediation, and Risk Technical Elements

S&T Activity	S&T Projects Planned	Project Objectives	Project duration (fiscal years)	Hanford Funding (thousand dollars)	EMSP Funding (thousand dollars)
Monitoring			2001-2003	0	2,600
Identification, development, and deployment of improved environmental monitoring	1	Investigate and develop strategies and technologies for environmental monitoring of multiple media			
Remediation			2001-2003	0	5,300
Identification, development, and deployment of improved groundwater remediation strategies	2	Investigate the distribution of DNAPLs in the 200 Area and contaminant plumes in the 100 Area, and assist in the development of corrective strategies			
Risk			2001	150	0
General risk assessment	1	Develop methods to identify, involve, and build consensus among stakeholders on desired foci for the risk technical element			
Ecological risk assessment	11	Develop methods to predict the impacts of exposures to contaminants from the Hanford Site on selected species	2001-2003	9,900[a]	0
Economic risk assessment	5	Develop methods and data for predicting how human populations and economies will respond from potential exposures of environmental contamination at the Hanford Site	2001-2003	3,300[a]	0
Human health risk assessment	9	Develop methods to predict the impacts on humans of exposures to contaminants from the Hanford Site	2001-2003	8,900[a]	0
Sociocultural risk assessment	1	Develop a risk perception model for groups affected by Hanford Site contaminants	2002	600[a]	0

NOTE: DNAPL = dense nonaqueous phase liquid; EMSP = Environmental Management Science Program
[a]The Integration Project intends to seek funding from national S&T programs (e.g., from DOE Headquarters) for some of this work.
SOURCE: DOE, 2000a, Figure 4-1, Table 5-1

In the second project (Rem-2), contaminant plumes (particularly chromium, strontium, and tritium plumes) in the 100 Area will be investigated, and the results will assist in the development of a strategy for corrective actions. The planned work will include geophysical and geochemical studies to help select and deploy remediation technologies.

The S&T activity for the Monitoring Technical Element, *identification, development, and deployment of improved environmental monitoring,* involves one project (M-1) to develop technologies and strategies to monitor air, vadose zone, groundwater, river, and selected biota, especially after the active phase of site cleanup is completed.

Evaluation of Work Planned Under the Monitoring and Remediation Technical Elements

The monitoring and remediation projects were begun before S&T plans were developed, so it is difficult for the committee to judge the appropriateness or effectiveness of current work. The total planned funding for these technical elements is $2.6 million and $5.3 million, respectively (last column of Table 9.1), all of which is being provided to principal investigators through the Environmental Management Science Program (EMSP, see Chapter 3). The Integration Project has no plans at present to provide Hanford Site funding to these efforts.

Detailed S&T plans for these technical elements are still not complete, and detailed objectives and S&T projects are not yet defined. If the planned projects are executed appropriately, however, the broad objectives defined for the Remediation Technical Element may be attainable. Knowledge of the distribution of contaminant plumes obtained from the planned S&T work will feed other Department of Energy (DOE) programs in the Office of Science and Technology that deal with remediation technology selection and deployment.

There is a clear opportunity to develop and evaluate new geophysical and geochemical methods for plume detection and treatment under these technical elements. The current field testing of in situ treatment of chromium in 100 Area groundwater is noteworthy in this regard.[4] Additionally, there exists much prior knowledge on remediating

---

[4] The In Situ Redox Manipulation Treatability Test is occurring near the 100-D Area. It involves the installation of a line of injection wells into the groundwater to intercept a plume of chromium(VI) just east of the Columbia River. A solution of sodium dichromate is being injected into the groundwater through these wells to form a chromium barrier. The solution reduces iron(III) in the sediments to iron(II), which in turn undergoes a redox reaction with chromium(VI) and reduces it to insoluble chromium(III) that precipitates onto the sediment grains. The test was

carbon tetrachloride and other dense nonaqueous phase liquids (DNAPLs) at other sites that should be applicable to Hanford.

New treatment methods and better technologies for locating and defining plumes should have broad application throughout the site (see Figure 2.8, for example) and to other DOE sites as well. Knowledge of the quantities and distributions of contaminants in the subsurface is essential so that decisions can be made to contain, actively treat, or rely on natural attenuation processes.

Not only will monitoring methods have to be developed to meet site needs for unprecedented time periods after the active phase of cleanup is completed, but monitoring results will be critical in decisions resulting in transitions from "watchful waiting" (surveillance) to active containment or remediation. However, there is insufficient information provided in the Integration Project Roadmap (DOE, 2000a) to determine whether the S&T work planned under the Monitoring Technical Element addresses Hanford's most critical needs in this area.

Discussion

In the committee's view, there is disproportionately small emphasis in the S&T program on remediation—only two projects are planned (Table 9.1), and both focus on determination of contaminant distributions in groundwater plumes. Little is being done *within* the S&T program on remediation technologies per se and their potential applications to waste pits, cribs, and other disposal areas. Further, there is no Integration Project S&T work planned on barriers to isolate, contain, and treat contaminants that will remain in the subsurface after the active cleanup program is completed.

The committee views the development of effective long-term barriers as one of the most important S&T needs at the Hanford Site, and an earlier National Research Committee (NRC) also identified this need as significant for DOE's national cleanup program (DOE, 2000a). In fact, S&T to develop and deploy effective barriers at Hanford has been under way for the better part of the last decade.

Field research on surface barriers in the 200 Area was initiated in the early 1990s with construction of the Field Lysimeter Test Facility (Fayer et al., 1999), followed by construction of the Hanford Surface Barrier over an actual waste site in 1994 (DOE, 1999d). These two

---

initiated in 1997 with the installation of a 50-meter-long segment of wells, and initial tests look promising. Recently, however, some breakthrough of chromium(VI) has been detected, likely due to oxygenation of groundwater near the test wells. The cause of this oxygenation is under investigation.

Monitoring, Remediation, and Risk Technical Elements

facilities are unique, with the lysimeter providing site-specific information on fluid infiltration rates and amounts through different surface soil and vegetative covers and the surface barrier providing site-specific data on the effectiveness of a specially designed cover system (the Hanford Surface Barrier) in preventing fluid infiltration into a waste crib (Figure 9.1).

In 1996, a feasibility study (BHI, 1996) examined possible designs for capillary-evapotranspiration barriers. These included (1) the Hanford Surface Barrier, a thick, multilayer, long-term barrier intended for use on the most contaminated sites (e.g., cribs and trenches; see Figure 9.1); (2) a standard Resource Conservation and Recovery Act (RCRA) C (hazardous waste) barrier; (3) a modified RCRA C barrier, to be used for

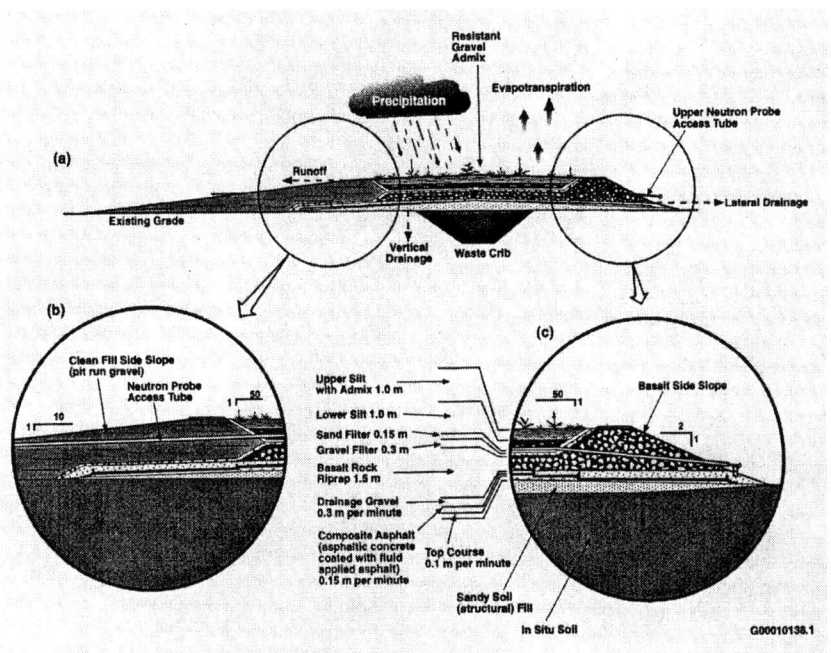

Figure 9.1 Cross-sectional view of the Hanford Barrier. SOURCE: DOE, 1998d, Figures 1.2, 1.3, 1.4.

low-level and mixed (radioactive and chemical) waste sites; and (4) a modified RCRA D (municipal waste) barrier for municipal and low-level waste sites without hazardous chemicals.

Some additional work on surface barriers is planned by DOE and Hanford, as noted below:[5]

- The Office of Science and Technology is initiating a three-year project in fiscal year 2001 on the Hanford Surface Barrier to (1) compare geophysical techniques, including ground-penetrating radar, for monitoring seasonal changes in moisture storage in the top two meters at scales larger than can be detected using neutron probe methods;[6] (2) determine the as-built water storage capacity of the barrier; and (3) validate the performance of the asphalt layer at the base of the barrier using chemical tracers.

- A study of the RCRA C barrier is also planned to begin in fiscal year 2001, with identification of local soil materials for use in its construction, followed by construction of a prototype barrier in the 200 Area in 2002 and performance monitoring thereafter. One of the Hanford Site's core projects, the 200 Area Remedial Action Project (see Chapter 3), is carrying out this work. Funding for the construction and monitoring phases is to be provided by Hanford and possibly by the Office of Science and Technology.

Detailed monitoring of properly constructed RCRA C (hazardous-toxic waste) and RCRA D (municipal waste) barriers has shown cover and liner systems to be very effective in preventing infiltration of water and escape of leachate. Some areas of the Hanford Site will remain too contaminated for any use other than waste disposal and containment for a very long time—for example, the 200 Area tank farms and disposal facilities like the Environmental Restoration Disposal Facility (see Chapter 2). Concentration and containment of waste using barriers with careful monitoring over time are environmentally defensible strategies.

Continued infiltration of wastes into the vadose zone in the 200 Area from tanks, cribs, waste ditches, and ponds may exacerbate the site remediation problems, as may the continued migration of contaminants away from these locations. Characterization data on the vadose zone (Chapter 5), combined with a fully functioning System Assessment

---

[5]This information was provided in a teleconference call with Integration Project and Pacific Northwest National Laboratory staff on October 3, 2000.

[6]Horizontal tubes were built into the Hanford Surface Barrier so that neutron probes could be inserted to measure moisture storage in the barrier materials (see Figure 9.1). Such methods provide only near-field moisture values—that is, values in the vicinity of the probe tubes.

Capability (SAC; see Chapter 4), can be used to illustrate how barriers over some or all of the source areas could reduce contaminant migration and plume development. Furthermore, use of surface and in-ground barrier systems can serve the additional purpose of containment, both temporary and long term. As new waste treatment technologies become available in the future, they could be used to treat and mitigate the hazards from wastes contained within them.

Interim barriers could find wide application across the site, for example, as barriers to reduce water infiltration in and around tank farms that have leaked waste,[7] or as subsurface barriers to reduce contaminant leakage into the vadose zone during single-shell tank waste retrieval operations. Access restrictions to and additional loading on the 200 Area tank farms have been cited to the committee as reasons for not placing interim barriers over them to reduce infiltration. However, there are options to avoid these problems; for example, tent-type structures that could easily be constructed over the tank farms, with grading to intercept surface runoff. S&T can play an important role in investigating and testing the feasibility of such strategies.

Perhaps the past and planned studies cited above account for the lack of a major S&T program on barriers as part of the Integration Project. Nonetheless, the committee believes that more research is needed on the development, deployment, and evaluation of interim, long-term, and reactive barrier systems. Recent research on new technologies for site remediation using chemically reactive barriers shows great promise. The development of reactive barrier systems for the unique site and waste conditions at Hanford could be an especially fruitful area of research.

The committee makes the following recommendations with respect to barrier S&T:

- **S&T research on the feasibility, effectiveness, and long-term performance of surface barriers should be expanded to include other barrier types and materials for prevention of surface water infiltration into tank farms and other regions containing high concentrations of buried and spilled waste.**

- **S&T should be undertaken to assess the potential for using vertical and inclined cutoff barriers and reactive barriers as part of "interim" waste containment systems, which can provide containment for one to a few decades, as well as "permanent" waste**

---

[7]At the request of DOE, a TechCon Forum was held in Richland, Washington, on May 4-6, 1999 to discuss approaches for reducing water infiltration around the Hanford tanks. The forum concluded that many technologies were already available or could be developed and deployed at the site (TechCon, 1999a,1999b).

containment systems that are designed to last for much longer periods.

- S&T should also be undertaken to develop reactive barriers that can be used to treat or immobilize radionuclide and chemical contaminants of concern in Hanford Site groundwater.

Since the development of new and improved barriers would likely find wide application across the DOE complex, much of the needed S&T work might be done in cooperation with other DOE programs. The focus of S&T at Hanford might be to adapt technologies developed elsewhere to the needs and environmental conditions at the site and perform pilot demonstrations.

The need for more and better methods for site characterization and monitoring has been cited throughout this report (see especially Chapters 5, 6, and 8). Monitoring is now being used at the site to detect contaminant transport in the environment, especially the groundwater and river (PNNL, 1999), and monitoring will no doubt find widespread use to assess the efficacy of future containment and cleanup actions. Monitoring utilizes many of the same strategies and tools as characterization and, in fact, is often piggybacked on,[8] or iterated with, characterization efforts.

Thus, as characterization capabilities are improved through Integration Project S&T efforts (see Chapter 5), there will be parallel opportunities to improve monitoring capabilities, especially through the use of minimally invasive or noninvasive approaches based on new detection, data collection (including improved statistical strategies for sampling), data transmission, data processing, and information display technologies. Such improvements are especially needed for monitoring the vadose zone, which will contain most of the waste and contamination to be left in place at the site when DOE's cleanup program is completed, and where contamination is especially difficult to detect because of subsurface heterogeneity and transient hydrologic conditions.

The development of new and improved monitoring strategies and capabilities is a difficult technical challenge, and it will likely take many years of effort to make useful progress. Consequently, S&T on monitoring strategies and tools must begin now so that useful results will be available before the initiation of large-scale remediation and containment activities at the site. **Therefore, the committee recommends that Integration Project S&T on new and improved strategies and technologies for monitoring the vadose zone be expanded. As part of this work, the Integration Project should assess what monitoring capabilities will be needed in the future at the site, based on cleanup decisions to be**

---

[8]For example, boreholes drilled to obtain data on subsurface characteristics may subsequently be used for groundwater monitoring.

made and likely end-state scenarios, so that the S&T work can be properly planned and prioritized.

Since the development of new and improved monitoring capabilities would likely find wide application across the DOE complex, much of the needed S&T work might be done in cooperation with other DOE programs. The focus of S&T at Hanford might be to adapt technologies developed elsewhere to the needs and environmental conditions at the site and perform pilot demonstrations.

RISK TECHNICAL ELEMENT

The Risk Technical Element has one general S&T activity on stakeholder involvement and four S&T activities focused on ecological, human health, economic, and sociocultural issues (Table 9.1). In general, the objective of the planned S&T work is to reduce uncertainties associated with risk assessments so that less conservative assumptions and models can be used. The approach taken is described in Appendix B and Table 4-1 in the Integration Project Roadmap (DOE, 2000a) and also in DOE (2000d).

1. *General risk assessment.* This activity includes one project (R-1) to develop methods to identify, develop, and build consensus among Hanford stakeholders on areas of focus for the Risk Technical Element.

2. *Ecological risk assessment.* This activity includes 11 projects (R-2 through R-12) to address technical gaps in ecological risk issues relevant to the Hanford Site. One project, described as contributing to better understanding of problem formulations, is designed to explore the relationship between impacts on individual entities and effects considered on a community level. Three exposure-related projects are described; these include the effects of multiple exposure pathways on uptake, uptake factors for plants and benthic species, and bioavailability. Six projects are defined that address ecological effects. These include a study of continuous (in contrast to acute) toxicological response, extrapolations across end points (such as population-level and individual-level responses) and taxa, toxicity by mode of action, and adaptive response. Finally, a risk characterization project is proposed in which a "weight-of-evidence" approach, as used for human health risk assessment, would be applied to the integration of data on ecological effects.

3. *Economic risk assessment.* This activity includes five projects (R-13 through R-17) to assess recreational use patterns for the Columbia River that can be used to define human exposure scenarios, assessment

of how people might respond to risk information of various types, preferences regarding ecological scenarios, and study of population mobility and benefits transfer.

4. *Human health risk assessment.* This activity includes nine projects (R-18 through R-26) that address bioavailability, food-chain transfer factors, biomarkers, exposure pathways-factors, variability in exposures, toxicokinetics, treatment of uncertainties in cancer slope factors, and characterization of multiple health end points. Several of these projects (food-chain transfer factors and exposure pathways) would address risk assessment issues relevant to more accurate characterization of exposures of Native Americans.

5. *Sociocultural risk assessment.* This activity consists of one project (R-27) to model risk knowledge, in which cultural experts would attempt to convert information on concentrations of contaminants in environmental media into impacts on cultural values.

Evaluation of Work Planned Under the Risk Technical Element

As shown in Table 9.1, work on some of these projects is under way and will be completed in fiscal year 2003. The total planned funding for this technical element is about $22.8 million (second-to-last column of Table 9.1). These budget estimates are taken from the Integration Project Roadmap (DOE, 2000a), which was published before the more detailed draft *Risk Assessment Science and Technology Plan* (DOE, 2000d) was issued. DOE (2000d) does not include budget information, so it is not clear to the committee whether the budget has evolved along with the technical project plans.

In comparison to DOE (2000a), the project descriptions in DOE (2000d) indicate that substantial progress has been made in defining risk-related projects concerning issues at the Hanford Site. A majority of the projects in the human health risk area apply to generic (rather than site-specific) risk assessment issues (e.g., R18-R20, R22, R23). These generic risk assessment issues have been recognized as important by many organizations and agencies, and technical advances in these areas would lead to reductions of uncertainty in risk estimates. However, these technical issues can have a strong influence on how environmental health risks are regulated and, for this reason, are of active interest to the Environmental Protection Agency (EPA) and other agencies. Additional comments on this point are offered later in this analysis.

Despite improvements in the planned risk project descriptions in DOE (2000d), many are still general and vague, such that detailed review

Monitoring, Remediation, and Risk Technical Elements 135

comments cannot be provided. For example, the description of the biomarkers-of-exposure project does not identify specific chemicals or technical approaches, nor does it provide any comparison with the current capabilities (e.g., sensitivity of measures of chromosome aberrations to ionizing radiation) relative to exposures of interest at the Hanford Site.

Can the objectives of this work be achieved?

The human health risk projects that involve factors specific to the Hanford Site are achievable. These projects (bioavailability, food-chain transfer factors, exposure pathways, and spatiotemporal variation in exposure) involve measurements and analyses that have been performed successfully in other contexts.

Other aspects of the risk element are more ambitious and seem unlikely to be achieved except over extended periods of time. The determination of health and ecosystem risks from complex mixtures is an example of such an issue; this has been studied by various government and research agencies for many years. Some of the scientific issues listed for the human health risk component are issues in which there is substantial regulatory involvement (determination of cancer slope factors, for example).

In general, regulators require the use of published guidance such as that found in EPA's Integrated Risk Information System (IRIS) for cancer slope factors. Although EPA periodically updates the IRIS values and will consider submitted information when it does so, this is not an area in which the Integration Project should expect large changes from the status quo. The only exception may be for rarely encountered substances for which the toxicological information base is poor. The ecosystem risk analysis is similarly hampered by a dependence on exposure-concentration-effect concentration pairs that are unknown or poorly defined for multiple stressors.

In the committee's judgment, these more ambitious proposed health science and ecosystem impact analysis activities would make more sense as components of long-term research supported by DOE Headquarters or other agencies such as EPA than as S&T under the Hanford Integration Project. To be useful at Hanford, any scientific advances in human health risk assessment would first have to be accepted by national and international scientific bodies, and then adopted by EPA. The committee notes that although a number of these ambitious projects appear in the S&T plan (DOE, 2000a, Table 4-1), they have not yet been funded under the Integration Project.

Does the planned work represent new science?

The achievable aspects of the work, such as a refined characterization of Native American diets and other exposure factors, would improve the quality and applicability of site risk assessments. Some of the more ambitious human health and ecosystem studies would represent new science if successful, but as noted above, the committee's view is that such work would be a better fit elsewhere in DOE or other agencies. The planned ecosystem risk analysis approach does not represent new science, and the planned studies to begin defining dose-response relationships for Columbia River flora and fauna are necessary, but studies of this type are standard toxicological testing elements of risk assessment.

Can the planned work have an impact on cleanup decisions at the Hanford site?

To the extent that Hanford cleanup decisions will be based on avoiding unacceptable risks to human health and the environment, the committee judges that this work can be helpful in several respects. The ecological work to refine the understanding of food webs could allow for more robust assessments of the effect of contaminant seepage into the Columbia River, and there is a critical need to build ecosystem risk foundations based on a comprehensive understanding of ecosystem structure and function in the Columbia River. However, the committee's view is that, taken as a whole, the Risk Technical Element is less likely to impact site decisions than is S&T to better characterize the locations concentration, and speciation of existing contaminants (see Chapter 5).

Does the planned work address important issues?

The committee believes that the planned work does address important issues. The primary objective of most Comprehensive Environmental Response, Compensation, and Liability Act (CERCLA) activities is the protection of human health and the environment in the future. The CERCLA process also identifies community concerns as important to site decisions. The risk element directly addresses these issues.

Are there other concerns, comments, or suggestions that should be considered by the Integration Project in executing the planned work?

The committee's main concern with the planned work, as noted in the previous discussion, is that it addresses issues that have been studied by various government and research agencies for many years and have substantial regulatory involvement. The work planned under the auspices of the Integration Project's S&T program is modest in comparison to the scope and magnitude of work on these issues by other agencies and is therefore unlikely to make major contributions to resolving these issues. Consequently, these issues are more appropriate targets for long-term research programs funded by DOE Headquarters and other federal agencies. The Integration Project's main focus should be to maintain awareness of this work and to use new results as they become available. Most of these projects of questionable value have not yet been funded by the Integration Project and probably should not be funded in the future for the reasons given above.

Discussion

The Risk Technical Element includes some potentially important work to identify ecological impacts that may result from contaminant seepage into the Columbia River. In particular, the identification of locations where contaminant concentration and characteristics of exposure can have substantial effects on Columbia River flora and fauna is likely to be very useful in future versions of the SAC.

The committee's review of the Risk Technical Element revealed two important issues that are not being addressed by the S&T program at present: (1) the impacts of extreme events on the risk assessment and (2) the appropriateness of the time period selected for risk assessment. The committee comments on these in the following paragraphs.

Rare but high-magnitude (also known as "extreme") events such as fires, floods, and earthquakes are considered routinely in risk assessments. A large range fire, such as occurred at the Hanford Site in the summer of 2000, could have a substantial effect on contaminant releases by removing protective ground cover, which could lead to increased infiltration and surface runoff and erosion (see Figure 9.2). Other episodic events also could affect the movement of subsurface contaminants, such as climate changes that result in drought or increased precipitation. Floods have been an important erosional agent at the Hanford Site, as evidenced by the geological record of extreme flooding during the past 100,000 years (Sidebar 9.1).

Figure 9.2 July 12, 2000, SPOT image of the Hanford Site showing the extent of the June 2000 range fire. Dark regions on the photo are burned areas. Copyright CNES/2000.

SIDEBAR 9.1 Extreme Events at Hanford

The Hanford Site is subject to a number of "extreme" (high-magnitude, and low-frequency) events that could lead to substantial transport of contaminants into the environment. Although the probability of occurrence of such rare events in a single year is low, the consequences of these events could be much greater than those of the advective-dispersive transport mechanisms in the vadose zone and groundwater that are modeled in the SAC. This is especially true over the time scales (10^3-10^5 years) during which wastes will remain hazardous.

The Hanford Site is a fire-prone ecosystem, as evidenced by widespread blazes in 1984 and 2000. The 1984 fire burned more than 80,000 hectares—approximately half of the Hanford Site. The June 27-July 1, 2000, blaze burned a similar area and destroyed 11 structures near the south edge of the site. Such fires probably have a recurrence period of decades, and they could play an important role in contaminant transport at the site.

The Hanford Site is also vulnerable to different types of flooding events. For example, significant risk of mobilizing waste may exist from failures of pressurized water mains. Much of the water supply infrastructure at Hanford is beyond its design life, as was illustrated by the September 26, 1996, rupture of a 36-centimeter pipeline that released 2 million liters of water into the S-SX Tank Farms (Anderson and Soler, 2000). Catastrophic failures are detected readily, but noncatastrophic leaks are more difficult to detect, and instrumentation to assess such leaks is not in place. DOE is in the process of deactivating some of the old water lines in the 200 Area, but leakage from active water lines may continue to be a problem as the infrastructure ages.

At longer time scales, the Hanford site is vulnerable to catastrophic flooding. The geologic record clearly shows repeated inundations during glacial periods in the last 100,000 years (see Sidebar 2.1). During this time, floodwaters from glacial Lake Missoula repeatedly scoured the Hanford Site with up to 2,500 cubic kilometers of water—ten times the volume of present-day rivers on Earth—when its ice dam became buoyant and broke during periods of deglaciation. The most recent scouring event occurred about 15,000 years ago (Waitt, 1985).

> The SAC focuses primarily on the standard subsurface aqueous transport pathway for release of contaminants from the site, with emphasis on the isolation capabilities of a thick, dry vadose zone (Chapter 4; see especially Figure 4.1). However, even under current (interglacial) climatic conditions, intense rainfall occasionally saturates the land surface and generates runoff and attendant sediment transport. Moreover, if one seriously considers a longer time frame (100,000 years), other exposure pathways are likely to become important—in particular, the erosion, transport, and redeposition of contaminated sediments and emplaced waste materials.

Documents and presentations to the committee indicate that risk assessments to be made using the SAC are to be carried out for of 1,000 years following site closure, from 2050 to 3050 (see Chapter 4). The committee has two recommendations with respect to this period of analysis. **First, the committee recommends that the results of such calculations be reviewed to ensure that the analytical period includes the time of peak dose or risk.** Such a review could be undertaken under the auspices of the Risk Technical Element. For some toxic materials, the rate of transport through the subsurface may be so slow that peak concentrations at locations of concern, especially the Columbia River, could occur more than 1,000 years in the future. If peak risks occur beyond 1,000 years, then other assumptions in the SAC may also need to be reexamined, particularly the assumption of no climate change (see Table 4.1).

The second recommendation concerns the "status quo" assumption made in the SAC with respect to the continuing existence of dams along the Columbia River. As discussed in Chapter 4 (see especially Table 4.1), the SAC contains the assumption that Columbia River dams will continue to operate for 1,000 years following the assumed closure of the Hanford Site in 2050. No justification for this assumption is given in any of the documents that the committee has reviewed, and it seems on its face to be unrealistic. **The committee recommends that an evaluation be made of the sensitivity of SAC risk assessments to the assumed continuing existence of these dams.** Again, such a review could be undertaken under the auspices of the Risk Technical Element. The committee is not recommending that analyses be made of a broad array of alternative future states with respect to the dams, only that the potential effect on analytical results of changes to the dams be considered.

10
Improving S&T Program Effectiveness

The statement of task for this study (see Chapter 1) called for the committee to review the Integration Project's science and technology (S&T) program and recommend ways to improve its technical merit and relevance to cleanup decisions at Hanford and other Department of Energy (DOE) sites. Much of this task has been accomplished in Chapters 5 through 9, which review the S&T projects within the seven technical program elements. This concluding chapter is structured around a set of findings and recommendations that are offered to improve overall program effectiveness.

Finding: There is a *long-term* and continuing need for S&T to support cleanup and stewardship of the Hanford Site.

As discussed in Chapter 1, environmental cleanup at Hanford is slated to last until at least 2046 and to cost upward of $85 billion (DOE, 1998e).[1] Moreover, after this active phase of cleanup is complete, the federal government's stewardship responsibilities will last for centuries. Hanford cleanup schedules are being driven by regulatory agreements and decisions that are not necessarily compatible with S&T time lines. This has led to S&T prioritization that may be inappropriate from a research or process development perspective.

DOE, its regulators, and the public face some hard truths about Hanford Site cleanup: the knowledge and technology to address the most difficult problems at the site do not yet exist. Consequently, much of the waste and contamination that is now in the subsurface, especially in the 200 Area, will very likely remain there for the foreseeable future. In addition, completion of Hanford cleanup could add substantially to this contamination, for example, during retrieval of tank waste. Currently, the range of available end-state, cleanup, containment, and monitoring options is greatly limited because of these knowledge and technology gaps. Advances in knowledge and technology will not be possible without continuing investments in S&T.

Given the long lead times for many of the planned end-state and cleanup decisions at the Hanford Site, there is an opportunity now to undertake S&T that could substantially advance DOE's capabilities to address the site's most difficult waste and contamination problems,

---

[1]Life-cycle costs fully escalated to year of expenditure. These are DOE estimates and have not been reviewed or validated by the committee.

especially for subsurface characterization, remediation, long-term containment, and monitoring (see Chapters 5 through 9). Many of these advances will be enabled by scientific discoveries[2] outside DOE that will undoubtedly occur over those same time spans. Continuing investments in S&T by DOE can help ensure that future cleanup and stewardship programs can take full advantage of such discoveries.

The Integration Project has the potential to provide much of S&T needed to advance the Hanford cleanup program over the coming decades. Based on the planning documents reviewed by the committee, however, it is not clear whether DOE plans to maintain this project beyond 2004 (e.g., Table 3.1). Clearly, a long-term commitment by DOE to S&T at Hanford will be essential for the future success of the site's cleanup and stewardship efforts.

> **Finding: Given the technical and organizational complexity of the task, the Integration Project has made a good start in creating an S&T roadmap, defining and initiating an S&T program, and fulfilling the promise of its mission.**

Although the committee has identified weaknesses in the S&T program, as noted throughout this report, the committee is impressed that the Integration Project has, over the short period of its existence, been able to initiate S&T work on sensible, high-priority projects in spite of numerous organizational and funding challenges. As discussed in Chapters 2 and 3, the Integration Project's task to provide S&T for site cleanup decisions is complicated by the number of organizations involved, the lack of clear authority and ownership, the extensive coordination requirements, and the lack of clearly defined site futures and cleanup decisions. The Integration Project comprises staff from several major Hanford contractor organizations and two DOE offices. Much of its work is being carried out in coordination with five core projects at Hanford and with the Environmental Management Science Program (EMSP), which is operated out of DOE Headquarters. Further, the Integration Project controls a small fraction of the S&T funding that supports its mission (Table 3.1) and does not have authority over the other parties operating the site or performing other S&T-related activities. Despite of these organizational obstacles, work is getting done.

As discussed elsewhere in this chapter, S&T priorities and activities ideally would be determined through a top-down framework in which high-level goals—in this case, site end states and the key cleanup

---

[2]The spectacular advances that have occurred since the second world war in information, communication, computation, bioengineering, and materials S&T should continue and may even accelerate in the decades ahead.

Improving S&T Program Effectiveness

decisions needed to achieve them—are used to set S&T priorities and schedules. Unfortunately, this framework was not in place prior to the establishment of the Integration Project. Instead, the S&T program was established to meet incompletely defined cleanup goals and schedules, with no authority to compel cooperation from other organizations at the site on which the project was superimposed (see Chapter 3) and with no guarantee of adequate or sustained funding levels.

Against the background of these constraints, the Integration Project has created and begun implementation of S&T activities that, taken as a whole, address some of the important contamination problems at the site. Although the current research agenda does not map against a defined set of information needs for meeting future site cleanup goals and technical details on many of the research projects are lacking, the Integration Project does appear to have developed a research portfolio that focuses on some of the important knowledge gaps at the site.

The committee believes that there are at least two reasons for the Integration Project's initial success in executing its S&T program despite these obstacles. First, the Integration Project appears to have effective leadership from both DOE and site contractor organizations.[3] The staff with which the committee had regular contact during its study, particularly the project managers,[4] were competent and enthusiastic, appeared to have instilled a sense of mission within the Integration Project staff, and also appeared to have established cooperative working relationships with the other entities at the site that are crucial to the project's success. Indeed, the Integration Project appears to have had some success in breaking through the organizational barriers at Hanford to encourage a cooperative atmosphere in which staff identify with projects rather than contractor organizations.

Second, the Integration Project appears, at present, to have the support of DOE Headquarters and Hanford Site management. For example, DOE Headquarters has provided additional direct financial support to the Integration project through the EMSP, including $1 million to support workshops to bring principal investigators to Hanford to interact with problem holders (see Chapter 3).

Finding: Although the S&T program has made a good start, its success is by no means guaranteed. Improvements are needed in the processes used to

---

[3] This statement is based on the committee's interactions with Integration Project management and staff at its six information-gathering meetings as well as limited interactions by telephone and e-mail outside of those meetings.

[4] Particularly Mike Thompson (DOE), Michael Graham (Bechtel Hanford, Inc.), and Mark Freshley and John Zachara (Pacific Northwest National Laboratory).

identify S&T priorities and to select, support, and manage S&T projects.

The Hanford remediation and stewardship project is one of the most complex and largest environmental projects ever undertaken. It involves numerous interacting cleanup projects planned over a period of about five decades, each of which will potentially have numerous and distinct S&T needs. In fact, the number of identified S&T needs is in the thousands (DOE, 2000b).[5,6] Only a small fraction of these needs are being pursued currently under the Integration Project S&T program or other programs such as the EMSP, largely due to time and funding constraints.

Therefore, a prioritization system is needed to identify those knowledge gaps that, if addressed successfully, could best advance the Hanford cleanup effort. Although some prioritization takes place every year at budget time and other processes exist to screen site needs on a regular basis,[7] there does not appear to be a formal and uniform prioritization system in place with specific criteria or guidelines that assign every S&T task at Hanford a priority ranking or number. This is true as well for the Integration Project S&T program. Given the lack of well-defined end states and cleanup decisions to be made at the site, the multiple organizations involved, and funding constraints, it is *essential* that an effective prioritization system be implemented to maximize the effectiveness of the S&T effort.

Recent efforts on the part of Hanford Site management to better define end states (DOE, 2000i) represent, in the committee's view, a welcome step forward in the cleanup program. Nevertheless, in the absence of well-defined end states, the Hanford cleanup program appears to operate on the philosophy that is better to take a step in approximately the right direction than to know exactly where it is going. The S&T program appears to be operating under the same philosophy. *This step-at-a-time approach to S&T may be useful during the early stages of cleanup when major knowledge gaps are easier to identify, but this approach probably will not work as well as the cleanup program matures and a long-term stewardship program is initiated and*

---

[5]The committee has not reviewed all of these needs to determine their relevance to site remediation or cleanup decisions.

[6]Hanford is developing another report entitled *Hanford Site Cleanup: Challenges and Opportunities for Science and Technology* that may contain additional needs. This report had not been released by the time the committee's report went to review, and the committee has not had an opportunity to review it.

[7]For example, the Hanford Site Technology Coordination Group collects and screens the S&T needs before they are sent to DOE Headquarters for selective contract awards.

implemented. In particular, this approach will make it difficult to uncover long-term research needs, which are not easily identified, even in well-planned programs.

The guiding principles for a useful S&T prioritization system are fairly straightforward: The S&T performed in support of Hanford cleanup should be relevant, should examine the best options of applicable alternatives, and should be cost-effective. At least three conditions must be satisfied to ensure that these requirements are met:

1. the critical decisions required to complete site cleanup must be defined;
2. the gaps in knowledge required to support such decisions must be identified; and
3. candidate S&T projects must be designed specifically to fill the identified knowledge gaps.

Once identified, of course, projects must be reviewed periodically to ensure that they continue to be applicable and are making appropriate progress. These points are addressed in more detail elsewhere in this chapter.

One of the most important criteria to be used in the prioritization system is the degree to which the S&T project contributes to the reduction in overall environmental risk and uncertainty (Sidebar 10.1) of a particular decision. In some S&T projects, uncertainty is the dominant issue. The degree to which the outcome of a particular S&T project is likely to reduce overall uncertainty must also be coupled with an economic analysis that compares the relative cost of the projects with the cost of proceeding with existing knowledge or the cost of reaching an incorrect decision if the project is not conducted. In addition to the uncertainties regarding the site and the future decisions that will be made with respect to site environmental risks, there are also technical risks with any individual S&T research activity or with portfolios of S&T activities. The fact that the degree of success of S&T activities is uncertain suggests that such project risks also be considered in the assignment of S&T priorities. This will facilitate determination of a project's cost-effectiveness.

The process described above can be applied in a straightforward manner when all of the cleanup decisions and data gaps have been predefined. When this is not the case, the process frequently can be applied in an iterative fashion. Intermediate cleanup goals and end states can serve as the basis for defining S&T needs, at least to support near-term work. As more knowledge is gained, the decision logic can be refined so that previously unrecognized data gaps may become apparent and the relative importance of previously identified data gaps may change. The S&T program can be refocused accordingly.

SIDEBAR 10.1: Uncertainty

Uncertainty can be defined as a lack of precise knowledge as to what the truth is, whether qualitative or quantitative (NRC, 1994b).

In the framework of risk management, it is useful to think about two types of uncertainty: *stochastic uncertainty*, which is caused by random variability in a process or phenomenon, and *state-of-knowledge uncertainty*, which results from a lack of precise or complete information about the processes or phenomena involved or their interactions. Conceptual models, described in mathematical terms, are used often to describe the physical behavior of the processes; phenomena of interest; and the components, subsystems, and systems involved and to calculate associated risks and uncertainties.

There are various sources of uncertainty. For example, when a particular property (parameter) is to be determined by measurement, estimates of the value of that parameter will be subject to the error that is inherent to the measurement process. Measurement uncertainty or error is typically, but not always, among the smallest contributions to the overall uncertainty in the value of a parameter.

If the parameter of interest represents a property of a large volume of material or a property that varies over time, estimates of the value of such a parameter also will be subject to sampling error and heterogeneity. Heterogeneity is the degree to which the parameter varies spatially or temporally over the volume or time period of interest.

Sampling error derives from the degree to which collection of a finite number of samples adequately represents the entire volume or time period of interest. This, in turn, depends on both the degree to which each individual sample is representative of the location from which it was collected and the design of the overall arrangement of spatial or temporal locations from which the set of samples are collected. Depending on the degree of heterogeneity in the parameter of interest, the design of the arrangement of samples collected, and the number of samples collected, the uncertainty contributed by these sources can sometimes be substantial.

Uncertainty introduced by lack of knowledge can be difficult to quantify. This error is introduced when there is an incomplete understanding of the nature of the processes or parameters of interest or of the factors that affect them. Substantial error can be introduced, for example, if the conceptual model for the system of interest (e.g.,

Improving S&T Program Effectiveness

> flow through the vadose zone) is incorrect or overly simplified so that important processes are neglected.
>
> For environmental problems, uncertainty typically derives from several sources (Capel and Larson, 2001):
>
> • uncertainties of input parameters, for example, locations and levels of contaminants and their variation with time;
>
> • uncertainties resulting from inadequate modeling of physical processes and phenomena, simplifying assumptions, or incomplete descriptions of the system, subsystem, or components;
>
> • uncertainties in results from experiments, including measurement or sampling errors, differences from actual conditions, and scaling distortions;
>
> • uncertainties and biases resulting from limited data; and incomplete understanding of the factors affecting the parameter of interest; and
>
> • completeness uncertainty, which accounts for whether all of the significant phenomena, processes, interactions, couplings, and events are considered.
>
> Once these various uncertainties have been determined, they must be combined into estimate overall uncertainty in a justifiable manner.

The committee did not observe the direct use of this risk- and uncertainty-based prioritization approach in the Integration Project S&T program. The Integration Project has given relatively greater priority and funding to S&T on the vadose zone over the other technical elements, presumably in recognition of the greater uncertainties in vadose zone contamination and fate and transport processes. As noted elsewhere in this report, the committee concurs with this prioritization. The committee believes that it would be useful, in an effort of this size and complexity, to systematically seek to identify the uncertainties that are most important to end-state and cleanup decisions at the site and to identify and select S&T projects that would most reduce those uncertainties to enable sound decisions to be made. To this end, systems-based analyses such as the System Assessment Capability (SAC) could be a useful tool for setting research priorities in the S&T program.

Recommendation: The Integration Project should develop and implement a system for planning and prioritizing its S&T activities to provide the information that Hanford Site management will need

to make sound and durable cleanup and stewardship decisions. An example of such a system is given in Sidebar 10.2.

Once the projects to be initiated are established, "owners" should be identified and held accountable for progress and costs. Successful management structures usually have clear lines of authority and accountability, and many organizations vest authority and accountability in a single centralized entity. In the case of the Integration Project and its S&T program, designating one person in charge of S&T who has outstanding technical and managerial skills and who reports to the Integration Project manager could improve the effectiveness of the program. The current structure does not appear to provide this clear management responsibility.[8]

> **Recommendation: The Integration Project should review its organization to ensure that ownership, authority, and accountability for the S&T program are clearly defined and assigned. Given the number of organizations involved in S&T and cleanup activities at the Hanford Site, help from DOE management above the level of the Integration Project may be needed to carry out this recommendation.**

Once priorities are established, the needed S&T work is carried out through a set of individual projects. As discussed in the previous chapters of this report, some of these projects are developed and supported by the Integration Project, whereas others are developed and supported through Hanford core projects or DOE Headquarters programs such as the EMSP. The S&T projects performed in support of the Hanford cleanup must satisfy the same criteria discussed above for the S&T program: they should be relevant, represent the best options of applicable alternatives, and be cost-effective. To ensure that management decisions are well informed, at least two conditions must be met:

1. The projects must be well documented, particularly with respect to objectives, technical study designs, work plans, products, schedules, and costs.

---

[8]At its March 28-30, 2001 meeting, the committee was informed by Michael Graham (Bechtel Hanford, Inc.) that management of the Integration Project may be transferred from Bechtel to another contractor in June 2002. The committee does not have enough information to determine whether such a transfer would resolve the management structure problem discussed in this section.

SIDEBAR 10.2: S&T Planning and Prioritization Tools

Several planning tools are available to guide the design of cost-effective S&T programs to provide timely and relevant information needed to reduce scientific uncertainty for sound cleanup, end-state, and land use decisions. One of these, the Data Quality Objectives (DQO) Process, is described here.

The DQO process (EPA, 2000a, 2000b) is a planning tool to facilitate more efficient and cost-effective designs of field investigations and to support improved decision making with reduced decision errors. The process promotes a comprehensive and systematic approach to problem solving and was developed for environmental projects where it is frequently necessary to make decisions in the face of substantial uncertainty (NRC, 1994b).

During planning of an environmental field investigation, the primary question that must be evaluated is whether reducing uncertainty will actually reduce the chance of making an incorrect decision. Depending on the manner in which the decision is to be determined and the nature of the uncertainties involved, reducing those uncertainties may not always reduce the decision error. In some cases, the cost to reduce the uncertainty may exceed the cost of making an incorrect decision in the first place. The DQO process is designed to force planners to address these issues so that they can identify the most cost-effective approaches for rendering decisions with acceptable accuracy.

The process itself consists the following steps, which are executed in a logical sequence: To determine the most cost-effective approach for solving a "problem" (e.g., remediating a site), it is first necessary to define that problem concisely. It is then necessary to reduce the problem to a series of one or more decisions about actions that define how a site will be modified to take it from its current state to a future, desired state. To the extent that options exist, these too should be identified. Each decision typically will be based on an evaluation of data, which must be specified concisely. If such data are not available or of good quality, then a "study" (e.g., an S&T project) may be required to fill the identified data gaps.

It is then necessary to evaluate the quality (accuracy and precision) of the data expected to be generated from the study to predict whether they will be sufficient to fill the targeted data gaps adequately. Finally, the cost of the study, along with some measure of the probability of success, must be weighed against alternate approaches, and also against the cost of the consequence of not obtaining the data, to determine whether the study is cost-effective.

> By explicitly addressing the handling of uncertainty and the cost trade-offs attendant on its control, the DQO process could, if used appropriately, substantially improve the selection and design of S&T studies that are performed in support of cleanup efforts at Hanford. Of course, the utility of the DQO process and other S&T planning tools will be only as good as the realism with which they are applied. These are not "off-the-shelf" tools, especially for complex applications such as Hanford, and they will require a great deal of careful thought and effort if they are to be applied successfully.
>
> The DQO approach is one of several possible approaches for priority setting. An excellent discussion of other systems-based approaches is given in NRC (1999b).

2. This documentation must be evaluated to ensure that the projects selected for funding are of high technical quality and are likely to meet S&T goals.

As noted in numerous places in this report, many of the current and planned S&T projects reviewed by the committee were poorly documented. Documentation on project objectives, technical study designs, work plans, and work products was frequently cryptic or unavailable.[9] Work schedules and cost information, when provided, generally were not useful for determining whether sufficient funding and time were being allowed for project objectives to be met.

There were, however, some clear exceptions to these generalizations. The EMSP projects, for example, were well documented, as were some of the projects supported under the Vadose Zone Technical Element, particularly the vadose zone transport field studies (see Chapter 6).

Because of the lack of documentation, many of the individual S&T projects were unreviewable by the committee. There was no basis to determine why some projects were included in the S&T roadmap or whether they would, if funded, meet the stated S&T goals. The committee believes that such projects would also be very difficult to manage for the same reasons.

Recommendation: The Integration Project should develop and implement guidelines for documenting

---

[9]The committee requested in writing the documentation for the S&T projects and was informed in writing that such documentation existed only for a small number of projects.

the objectives, technical study designs, work plans, work products, work schedules, and costs for its S&T projects. To this end, the Integration Project should consider and adapt, as appropriate, guidelines from other S&T programs such as the Environmental Management Science Program.

As noted in Chapters 1 and 3 of this report, one of the primary objectives of the Integration Project is to "[e]stablish an independent technical peer review" of the work under its purview. The work of this committee and the Integration Project Expert Panel (see Chapter 3) are manifestations of DOE's commitment to this objective. Other examples of this commitment include DOE-sponsored peer review (through DOE Headquarters) of EMSP projects supported under the Vadose Zone, Monitoring, and Remediation Technical Elements (see Chapters 6 and 9), as well as a peer review of the Hanford Site groundwater model (Gorelick et al., 1999).

The committee agrees with DOE that peer review should be an essential element of the Integration Project. Peer review can provide independent assessments of the technical merit and relevance of the proposed work, an opportunity for midcourse adjustments in project plans and/or experiments, and an assessment of the quality of the work that has been completed. Peer review also can provide valuable alternate perspectives to the project and can be an efficient means of alerting project staff to research efforts and progress outside DOE.[10]

Although DOE is committed in principle to peer review of Integration Project S&T, it is too early in the project to determine exactly how such reviews will be implemented, especially for individual projects.[11] The committee believes that there is likely to be a benefit to the Integration Project if peer review is applied in all aspects of the S&T program.

Recommendation: Peer review[12] should be used for program prioritization, selection of S&T projects to

---

[10]See also the recommendation of peer review of vadose zone transport field studies in Chapter 6.

[11]Projects supported by DOE Headquarters, for example the EMSP and other Office of Science and Technology projects, are routinely selected for funding on the basis of peer review.

[12]A peer review is a documented, critical review performed by "peers" (i.e., persons having technical expertise in part or all of the subject matter to be reviewed) who are independent of the work being reviewed. The peer's independence from the work being reviewed means that the peer was not involved as a participant, supervisor, or adviser in the work being reviewed and, to the extent practical, has sufficient freedom from funding considerations to ensure

be funded, and periodic assessments of multiyear projects to ensure that they continue to meet program objectives. To this end, Integration Project should consider and adapt, as appropriate, guidelines from other S&T programs—for example, DOE's Office of Science, DOE's Environmental Management Science Program, and the National Science Foundation.

Of course, once S&T projects are reviewed and selected, funding must be provided to carry out the proposed work. At present, the Integration Project funding has not been sufficient to support the selected projects due to reductions in planned budgets (see Table 3.1). In response to a question from the committee about the impact of funding reduction on the S&T program in fiscal year 2001, the Integration Project stated that

> [m]ore than 50 percent of the research planned will not be done as planned. This shortfall will impact the duration of the S&T effort and what will eventually be accomplished. Of the S&T research activities documented in Rev. 0 and Rev. 1 of the S&T roadmap ... several areas have not been funded, including significant portions of the Groundwater and Columbia River technical elements, and more recently, the Risk technical element. Within the other technical elements, the budget restrictions will result in less work being performed.

The committee has not performed a detailed analysis of the Integration Project's budget and does not have enough information to determine whether or not the current funding level is appropriate. The committee observes, however, that the current funding level ($4.6 million) is low relative to the magnitude of the current $1 billion plus annual cleanup effort at Hanford.

However, S&T is being carried out by other organizations at Hanford and DOE Headquarters, so the total investment in S&T is much greater than $4.6 million. S&T work is also being carried out by the core projects and the Office of Science and Technology at DOE Headquarters (see Chapter 3). However, this S&T work is not organized or reviewed on a system basis, and it is not clear how approval and funding decisions are

---

that the review is impartial (from USNRC, 1988, p. 2). A detailed discussion of peer review as applied to DOE science and technology programs is provided in NRC (1998).

prioritized across the Hanford Site or the Environmental Management (EM) Program.

Examination of Hanford Site and relevant EM S&T work on a system basis and its prioritization accordingly could be of great benefit to S&T planning and effectiveness, especially to determine whether the planned investments in Integration Project S&T are appropriate. Once this examination is completed, the adequacy of funding for Integration Project S&T can be better evaluated. Additionally, the appointment of S&T personnel to spearhead the S&T work for each critical system could enhance the coordination and effectiveness of that work.

Regardless of absolute funding levels, the lack of stable funding is impeding the Integration Project's ability to plan and execute its work. Delays in completing current and planned S&T work will delay the transfer of potentially important S&T results to the cleanup program.

Recommendation: The Integration Project should, with the help of EM as necessary, perform a system-based analysis of its funding needs for the S&T program once it develops the prioritization process recommended above.

References

Agnew, S.F. 1997. Hanford Tank Chemical and Radionuclide Inventories (HDW Model Rev. 4). Los Alamos, N.M.: Los Alamos National Laboratory. LA-UR-96-3860.

Agnew, S.F, and R.A. Corbin. 1998. Analysis of SX Farm Leak Histories—Historical Leak Model (HLM). Los Alamos, N.M.: Los Alamos National Laboratory.

Anderson, F.J., and L. Soler. 2000. Interim measures to limit the migration of radioactive contaminants through the vadose zone at Hanford's single shell tank farms [abstract]. EOS, Transactions American Geophysical Union 81(48): F413.

Bechtel Hanford, Inc. (BHI). 1996. Focused Feasibility Study of Engineered Barriers for Waste Management Units in the 200 Areas. Richland, Wash.: Bechtel Hanford, Inc. DOE/RL-93-33.

Bechtel Hanford, Inc. (BHI). 1999. Groundwater/Vadose Zone Integration Project: Preliminary System Assessment Capability Concepts for Architecture, Platform and Data Management, Richland, Wash.: Bechtel Hanford, Inc.

Becker, C.D. 1990. Aquatic Bioenvironmental Studies: The Hanford Experience 1944-1984. Studies in Environmental Science 39. Elsevier, Amsterdam.

Brevick, C.H. 1994. Historical Tank Content Estimate for the Northeast Quadrant of the Hanford 200 East Areas. Richland, Wash.: Westinghouse Hanford Company. WHC-SD-WM-ER-349 Rev 0.

Brevick, C.H. 1995a. Historical Tank Content Estimate for the Southwest Quadrant of the Hanford 200 West Area. Richland, Wash.: Westinghouse Hanford Company. WHC-SD-WM-ER-352 Rev 0.

Brevick, C.H. 1995b. Historical Tank Content Estimate for the Northwest Quadrant of the Hanford 200 West Area. Richland, Wash.: Westinghouse Hanford Company. WHC-SD-WM-ER-351 Rev 0.

Brevick, C.H. 1995c. Historical Tank Content Estimate for the Southeast Quadrant of the Hanford 200 East Areas. Richland, Wash.: Westinghouse Hanford Company. WHC-SD-WM-ER-350 Rev 0.

Buesseler, K.O., M.H. Dai, S. Pike, J.M. Kelley, and J.F Wacker. 2000. Speciation, Mobility and Fate of Actinide Element Isotopes in Groundwater. Presentation at EMSP Vadose Zone Principal Investigator Workshop, Nov. 28-Dec. 03, 2000.

Capel, P.D., and S.J. Larson. 2001. Effect of scale on the behavior of atrazine in surface waters. Environ. Sci. Technol. 35: 648-657.

References

Carroll, S., C.I. Steefel, and P. Zhao. 2000. Cesium transport in Hanford sediments: Integration of batch and column experiments [abstract]. EOS, Transactions American Geophysical Union, 81(48): F405.

Fayer, M.J., E.M. Murphy, J.L. Down, F.O. Khan, C.W. Lindenmeier, and B.N. Bjornstad. 1999. Recharge Data Package for the Immobilized Low-Activity Waste 2001 Performance Assessment. Richland, Wash.: Pacific Northwest National Laboratory. PNNL-13033.

Farris, W.T., B.A. Napier, T.A. Ikenberry, J.C. Simpson, and D.B. Shipler. 1994. Atmospheric Pathway Dosimetry Report, 1944-1992. Richland, Wash.: Battelle Pacific Northwest Laboratories. PNWD-2228 HEDR.

Flury, M., J. Mathison, and J. Harsh. 2000. In situ mobilization of colloids in Hanford sediments during tank leakage [abstract]. EOS, Transactions American Geophysical Union 81(48): F414-F415.

Frissell, C.A., W.J. Liss, C.E. Warren, and M.D. Hurley. 1986. A hierarchical framework for stream habitat classification: Viewing streams in a watershed context. Environmental Management 10: 199-214.

Gephart, R.E. 1999. A short history of plutonium production and nuclear waste generation, storage, and release at the Hanford Site. Richland, Wash.: Pacific Northwest National Laboratory. PNNL-SA-32152.

Gephart R.E., and R.E. Lundgren. 1998. Hanford Tank Cleanup: A Guide to Understanding the Technical Issues. Columbus, Ohio: Battelle Press. PNL-10773.

Gerber, M.S. 1992. Legend and Legacy: Fifty Years of Defense Production at the Hanford Site. Richland, Wash.: Westinghouse Hanford Company. WHC-MR-0293. Rev. 2.

Gorelick, S., C. Andrews, and J. Mercer. 1999. Report of the Peer Review Panel on the Proposed Hanford Site-Wide Groundwater Model. On-line. Available at http://terrassa.pnl.gov:2080/gwmodeling/peerreport.html. Accessed October 26, 2000.

Hanlon, B.M. 2000. Waste Tank Summary Report For Month Ending July 31, 2000. Richland, Wash.: CH2MHill Hanford Group, Inc. HNF-EP-0182 Rev.148.

Hartman, M.J., L.F Morasch, and W.D. Webber, eds., 2000. Hanford Site Groundwater Monitoring for Fiscal Year 1999. Richland, Wash.: Pacific Northwest National Laboratory. PNNL-13116.

Heeb, C.M. 1994. Radionuclide Releases to the Atmosphere from Hanford Operations, 1944-1972. Richland, Wash.: Pacific Northwest National Laboratory. PNWD-2222-HEDR.

Heeb, C.M., and D.J. Bates. 1994. Radionuclide Releases to the Columbia River from Hanford Operations, 1944-1971. Richland, Wash.: Battelle Pacific Northwest Laboratories. PNWD-2223-HEDR.

Kerans, B.L., and J.R. Karr. 1994. A benthic index of biotic integrity (B-IBI) for rivers of the Tennessee Valley. Ecological Applications 4: 768-785.

Kersting, A.B., D.W. Efurd, D.L. Finnegan, D.J. Rokop, D.K. Smith, and J.L. Thompson. 1999. Migration of plutonium in ground water at the Nevada Test Site. Nature 397(6714): 56-59.

Kincaid, C.T., P.W. Eslinger, W.E. Nichols, A.L. Bunn, R.W. Bryce, T.B. Miley, M.C. Richmond, S.F. Snyder, R.L. Aaberg. 2000. Groundwater/Vadose Zone Integration Project System Assessment Capability (Revision 0): Assessment Description, Requirements, Software Design, and Test Plan. Richland, Wash.: Bechtel Hanford, Inc. BHI-01365.

Meinzer O.E., and L.K. Wenzel. 1942. Movement of ground water and its relation to head, permeability, and storage: in Meinzer, O.E., ed., Hydrology: National Research Council, Physics of the Earth 9: 444-477. New York: McGraw-Hill Book Co., Inc.

Minshall, G.W. 1988. Stream ecosystem theory: a global perspective. J. N. Am. Benthol. Soc. 74(4): 263-288.

Myers, D.A., and G.W. Gee. 2000, Moisture distribution in Hanford's SX Tank Farm [abstract]. EOS, Transactions American Geophysical Union, 81(48): F386.

Napier, B.A. 1992. Determination of Radionuclides and Pathways Contributing to Cumulative Dose; Hanford Environmental Dose Reconstruction Project Dose Code Recovery Activities—Calculation 004. Richland, Wash.: Battelle Pacific Northwest Laboratories. BN-SA-3673 HEDR.

National Research Council (NRC). 1994a. The Hanford Environmental Dose Reconstruction Project: A Review of Four Documents. Washington, D.C.: National Academy Press.

National Research Council (NRC). 1994b. Science and Judgment in Risk Assessment. Washington, D.C.: National Academy Press.

National Research Council (NRC). 1995. A Review of Two Hanford Environmental Dose Reconstruction Project (HEDR) Dosimetry Reports: Columbia River Pathway and Atmospheric Pathway. Washington, D.C.: National Academy Press.

National Research Council (NRC). 1996. The Hanford Tanks: Environmental Impacts and Policy Choices. Washington, D.C.: National Academy Press.

National Research Council (NRC). 1997. Building an Effective Environmental Management Science Program: Final Assessment. Washington, D.C.: National Academy Press.

National Research Council (NRC). 1998. Peer Review in Environmental Technology Development Programs: The Department of Energy's Office of Science and Technology. Washington, D.C.: National Academy Press.

National Research Council (NRC). 1999a. An End State Methodology for Identifying Technology Needs for Environmental Management, with an Example from the Hanford Site Tanks. Washington, D.C.: National Academy Press.

National Research Council (NRC). 1999b. Environmental Cleanup at Navy Facilities: Risk-Based Methods. Washington, D.C.: National Academy Press.

National Research Council (NRC). 2000a. Research Needs in Subsurface Science. Washington, D.C.: National Academy Press.

National Research Council (NRC). 2000b. Hanford Thyroid Disease Study Draft Final Report. Washington, D.C.: National Academy Press.

National Research Council (NRC). 2000c. Long-Term Institutional Management of U.S. Department of Energy Legacy Waste Sites. Washington, D.C.: National Academy Press.

Pacific Northwest National Laboratory (PNNL). 1999. Hanford Site Environmental Report for Calendar Year 1998: Richland, Wash.: Pacific Northwest National Laboratory. PNNL-12088.

Pacific Northwest National Laboratory (PNNL). 2000a. Hanford Site Environmental Report for Calendar Year 1999: Richland, Wash.: Pacific Northwest National Laboratory. PNNL-12088.

Pacific Northwest National Laboratory (PNNL). 2000b. Answers to Questions Concerning the Hanford Site from National Research Council Committee Meeting #2, June 20-22. Richland Wash.: Pacific Northwest National Laboratory.

Rhodes, R. 1986. The Making of the Atomic Bomb. New York: Simon and Schuster.

Rickard, W.H., and R.H. Gray. 1995. The Hanford Reach of the Columbia River: A refuge for fish and riverine wildlife and plants in eastern Washington. Natural Areas Journal 15(1): 68-74.

Rickard, W.H., and D.G. Watson. 1985. Four decades of environmental change and their influence upon native wildlife and fish on the mid-Columbia River, Washington, USA. Environmental Conservation 12(3): 241-248.

Rohay, V.J. 2000. Performance Evaluation Report for Soil Vapor Extraction Operations at the Carbon Tetrachloride Site, February 1992 – September 1999. Richland, Wash.: Bechtel Hanford, Inc. BHI-00720 Rev. 4.

Rosenberg, D.M., and V.H. Resh, eds., 1993. Freshwater Biomonitoring and Benthic Macroinvertebrates. New York: Chapman and Hall.

Schaeffer, D.J., E.E. Herricks, and H.W. Kerster. 1988. Ecosystem Health: I. Measuring Ecosystem Health. Environ. Management 12(4): 445-456.

Schaeffer, D.J., and E.E. Herricks. 1993. Biological Monitors of Pollution, in M. Corn, ed., Handbook of Hazardous Materials. New York: Academic Press: 69-80.

Stonestrom, D.A. 1996. An overview of unsaturated-flow theory as applied to the phenomena of infiltration and drainage: in Stevens, P.R., and T.J. Nicholson, eds., Joint U.S. Geological Survey, U.S. Nuclear Regulatory Commission Workshop on Research Related to Low-Level Radioactive Waste Disposal: U.S. Geological Survey Water Resources Investigations Report 95-401: 83-90.

TechCon. 1999a. Forum Proceedings Volume 1, May 4-6, 1999. Reducing Water Infiltration Around Hanford Tanks. Hills Conference Center, Richland, Wash.

TechCon. 1999b. Forum Proceedings Volume 2, May 4-6, 1999. Reducing Water Infiltration Around Hanford Tanks. Hills Conference Center, Richland, Wash.

Thompson, J.N., O.J. Reichman, P.J. Morin, G.A. Polis, M.E. Power, R.W. Sterner, C.A. Couch, L. Gough, R. Holt, D.U. Hooper, F. Keesing, C.R. Lovell, R.T. Milne, M.C. Molles, D.W. Roberts, and S.Y. Strauss. 2001. Frontiers of Ecology. Bioscience 51(1):15-24.

U.S. Department of Energy (DOE).1992a. 200 East Groundwater Aggregate Area Management Study Report. Richland, Wash.: U.S. Department of Energy. DOE/RL-92-19.

U.S. Department of Energy (DOE).1992b. 200 West Groundwater Aggregate Area Management Study Report. Richland, Wash.: U.S. Department of Energy. DOE/RL-92-16.

U.S. Department of Energy (DOE).1996. SX Tank Farm Report: Grand Junction, Col.: U.S. Department of Energy. DOE/ID-12584-268.

U.S. Department of Energy (DOE). 1997a. History of the Plutonium Production Facilities at the Hanford Site Historic District, 1943 1990: Richland, Wash.: U.S. Department of Energy. DOE/RL-97-

1947. On-line. Available at http://www.hanford.gov/docs/rl-97-1047/index.htm. Accessed March 2, 2001.

U.S. Department of Energy (DOE). 1997b. Tank Waste Remediation System Vadose Zone Contamination Issue: Independent Expert Panel Status Report. Richland, Wash.: U.S. Department of Energy. DOE/RL-97-49 Rev. 0.

U.S. Department of Energy (DOE). 1997c. Waste Site Grouping for 200 Areas Soil Investigations. Richland, Wash.: U.S. Department of Energy. DOE/RL-96-81 Rev. 0.

U.S. Department of Energy (DOE). 1997d. Linking Legacies: Connecting the Cold War Nuclear Weapons Production Processes to Their Environmental Consequences. Washington, D.C.: U.S. Department of Energy. DOE/EM-0319.

U.S. Department of Energy (DOE). 1998a. Groundwater/Vadose Zone Integration Project: Background Information and State of Knowledge: Richland, Wash.: U.S. Department of Energy. DOE/RL-98-48 Vol. II Rev 0.

U.S. Department of Energy (DOE). 1998b. Accelerating Cleanup: Paths to Closure. Washington, D.C.: U.S. Department of Energy. DOE/EM-0362.

U.S. Department of Energy (DOE). 1998c. Management and Integration of Hanford Site Groundwater and Vadose Zone Activities. Richland, Wash.: U.S. Department of Energy. DOE/RL-98-03.

U.S. Department of Energy (DOE). 1998d. Groundwater/Vadose Zone Integration Project Specification. Richland, Wash.: U.S. Department of Energy. DOE/RL-98-48. Draft C.

U.S. Department of Energy (DOE). 1998e. Accelerating Cleanup: Paths to Closure, Hanford Site. Richland, Wash.: U.S. Department of Energy. DOE/RL-97-57. Rev. 0.

U.S. Department of Energy (DOE). 1998f. Hanford Site Historic District: Reactor Operations. Richland, Wash.: U.S. Department of Energy. DOE/RL-97-Internet-1047. On-line. Available at http://www.hanford.gov/docs/rl-97-1047/reactor_ops/contributions.htm. Accessed March 2, 2001.

U.S. Department of Energy (DOE). 1998g. Commercial Nuclear Fuel from U.S. and Russian Surplus Defense Inventories: Materials, Policies, and Market Effects. Washington, DC.: Energy Information Administration. DOE/EIA-0619.

U.S. Department of Energy (DOE). 1999a. Final Hanford Comprehensive Land-Use Plan Environmental Impact Statement: Richland, Wash.: U.S. Department of Energy. DOE/EIS-0222F.

U.S. Department of Energy (DOE). 1999b. Groundwater/Vadose Zone Integration Project Science and Technology Summary

Description. Richland, Wash.: U.S. Department of Energy. DOE/RL-98-48. Vol. III. Rev. 0.

U.S. Department of Energy (DOE). 1999c. Groundwater/Vadose Zone Integration Project Inventory Scoping Study for the System Assessment Capability. Richland, Wash.: U.S. Department of Energy.

U.S. Department of Energy (DOE). 1999d. 200-BP-1 Prototype Barrier Treatability Test Report. Richland, Wash.: U.S. Department of Energy. DOE/RL-99-11. Rev. 0.

U.S. Department of Energy (DOE). 1999e. Richland Environmental Restoration Project. 2000 Cost Account Scope Statement. Richland, Wash.: U.S. Department of Energy. DOE/RL-97-44 Vol. 3 Rev. 2.

U.S. Department of Energy (DOE). 1999f. From Cleanup to Stewardship. Washington, D.C.: U.S. Department of Energy. DOE/EM-0466.

U.S. Department of Energy (DOE). 2000a. Groundwater/Vadose Zone Integration Project Science and Technology Summary Description. Richland, Wash.: U.S. Department of Energy. DOE/RL-98-48. Vol. III. Rev. 1.

U.S. Department of Energy (DOE). 2000b. Hanford Science and Technology Needs Statements. Richland, Wash.: U.S. Department of Energy. DOE/RL-98-01 Rev. 2.

U.S. Department of Energy (DOE). 2000c. Hanford Site Groundwater/Vadose Zone Integration Project Semi-Annual Report, October 1999 - March 2000. Richland, Wash.: U.S. Department of Energy/Bechtel Hanford, Inc.

U.S. Department of Energy (DOE). 2000d. Risk Assessment Science and Technology Plan. Richland, Wash.: U.S. Department of Energy.

U.S. Department of Energy (DOE). 2000e. Fiscal Year 1999 Annual Summary Report for the 200-UP-ZP-1, and 100-NR-2 Pump-and-Treat Operations and Operable Units. Richland, Wash.: U.S. Department of Energy. DOE/RL-99-97. Rev. 0.

U.S. Department of Energy (DOE). 2000f. Hanford Site-Columbia River Corridor Cleanup (Draft). Richland, Wash.: U.S. Department of Energy. DOE/RL-2000-66.

U.S. Department of Energy (DOE). 2000g. Richland Environmental Restoration Project. 2001 Cost Account Plan. Richland, Wash.: U.S. Department of Energy. DOE/RL-97-44 Vol. 3 Rev. 3.

U.S. Department of Energy (DOF). 2000h. Buried Transuranic-Contaminated Waste Information for U.S. Department of Energy Facilities. Washington, D.C.: U.S. Department of Energy.

U.S. Department of Energy (DOE). 2000i. Hanford 2012: Accelerating Cleanup and Shrinking the Site. Richland, Wash.: U.S. Department of Energy.

U.S. Department of Energy (DOE). 2001. A Report to Congress on Long-Term Stewardship: Washington, D.C.: U.S. Department of Energy. DOE/EM-0563 (2 Vols.).

U.S. Environmental Protection Agency (EPA). 2000a. Guidance for the Data Quality Objectives Process. Washington, D.C.: Office of Environmental Information. EPA QA/G-4. EPA/600/R-96/055.

U.S. Environmental Protection Agency (EPA). 2000b. Guidance for the Data Quality Objectives Process for Hazardous Waste Sites. Washington, D.C.: Office of Environmental Information. EPA QA/G-4HW Final. EPA/600/R-00/007.

U.S. Energy Research and Development Administration (ERDA). 1975. Final Environmental Statement, Waste Management Operations, Hanford Reservation. Richland, Wash.: U.S. Environmental Research and Development Administration. ERDA-1538 (2 vols.).

U.S. General Accounting Office (GAO). 1992. Nuclear Waste: Improvements Needed in Monitoring Contaminants in Hanford Soils. Washington, D.C.: U.S. General Accounting Office. RCED-92-149.

U.S. General Accounting Office (GAO). 1998. Nuclear Waste: Understanding of Waste Migration at Hanford is Inadequate for Key Decisions: Washington, D.C.: U.S. General Accounting Office. RCED-98-80.

U.S. Nuclear Regulatory Commission (USNRC), 1998. Peer Review for High-Level Nuclear Waste Repositories: Generic Technical Position. Washington, D.C.: U.S. Nuclear Regulatory Commission. NUREG-1297.

Vaughan, B.E., and J.L. Hebling. 1975. A bibliography of environmental research: Ecosystems Department 1952-1975. Richland, Wash.: Battelle Pacific Northwest Laboratories. BNWL-SA-4655.

Waite, J.L. 1991. Tank Wastes Discharged Directly to the Soil at the Hanford Site: Richland, Wash.: U.S. Department of Energy. Office of Environmental Restoration. WHC-MR-0227.

Waitt, R.B., Jr. 1985. Case for periodic, colossal jökulhlaups from Pleistocene glacial Lake Missoula. Geological Society of America Bulletin 96:1271-1286.

Wan, J., T.K. Tokunaga, E. Saiz, and K. Olson. 2000. Colloid formation and permeability reduction during infiltration of waste tank solutions into Hanford sediments [abstract]. EOS, Transactions American Geophysical Union, 81(48): F405-406.

Ward, J. V. 1989.The four-dimensional nature of lotic ecosystems. J. N. Am. Benthol. Soc. 8(2): 2-8.

Williams R.N., P.A. Bisson, C.C. Coutant, D. Goodman, J. Lichatowich, W. Liss, L. McDonald, P. Mundy, B. Riddell, R.A. Whitney. 1998. Recommendations for stable spring flows in the Hanford Reach during the time when juvenile Fall Chinook are present each spring. Portland, Or.: Independent Scientific Advisory Board, Northwest Power Planning Council, National Marine Fisheries Service. On-line. Available at http://www.nwppc.aa.psiweb.com:80/98-5hanf.htm. Accessed July 27, 2001.

Wodrich, D. 1991. Historical Perspective of Radioactivity Contaminated Liquid and Solid Wastes Discharged or Buried in the Ground at Hanford: Richland, Wash.: Westinghouse Hanford Company. TRAC-0151-VA.

Zorpette, G. 1996. Hanford's nuclear wasteland. Scientific American (May): 88-97.

A
Biographical Sketches

CHRIS G. WHIPPLE (*Chair*, NAE) is a principal in ENVIRON International Corporation in Emeryville, California. His professional interests are in risk assessment, and he has consulted widely in this field for private clients and government agencies. Prior to joining ENVIRON, he worked for ICF Kaiser Engineers (1990-2000) and the Electric Power Research Institute (1974-1990). He served on the National Research Council's (NRC's) Board on Radioactive Waste Management (BRWM) from 1985 to 1995, and as its chair from 1992 to March 1995. He is a past-president and fellow of the Society for Risk Analysis. He holds a B.S. degree from Purdue University and M.S. and Ph.D. degrees in engineering science from the California Institute of Technology.

D. WAYNE BERMAN is president of Aeolus, Inc., in Albany, California, and has more than 20 years' experience in solving complex environmental problems for a variety of government and private clients. He has extensive experience in chemical fate and transport; chemical process analysis; sampling and analytical method development; data quality analysis; data quality objectives development; risk assessment; and regulatory compliance. He holds a B.S. degree in chemistry from Muhlenburg College and a Ph.D. degree in physical chemistry from the California Institute of Technology. Dr. Berman is a member of the New York Academy of Sciences, American Chemical Society, American Association for the Advancement of Science, and American Society for Testing and Materials.

SUE B. CLARK is the Meyer Distinguished Associate Professor at Washington State University in Pullman, Washington. Dr. Clark's current research focus is on the oxidation-reduction chemistry of plutonium in natural aquatic systems. She served previously on the BRWM's Committee on the Waste Isolation Pilot Plant. She holds a B.S. degree from Lander College and M.S. and Ph.D. degrees in inorganic and radiochemistry from Florida State University. Prior to joining Washington State University, she worked as an assistant research ecologist at the University of Georgia's Savannah River Ecology Laboratory and as a research assistant at Katholieke University te Leuven in Belgium. Dr. Clark has received the Westinghouse Savannah River Company's Total Quality Achievement Award. She is a member of Sigma Xi, the Scientific Research Society, and the American Chemical Society.

JOHN C. FOUNTAIN is professor and chair of the department of geology at the University at Buffalo, State University of New York. Dr. Fountain's research has focused on contaminant hydrology, specifically aquifer remediation and characterization of fractured rock aquifers. He has served on several BRWM study committees, including the Committee on Technologies for Cleanup of Subsurface Contaminants in the Department of Energy (DOE) Weapons Complex. Dr. Fountain holds a B.S. degree in chemistry from California Polytechnic State University, and M.A. and Ph.D. degrees in geology from the University of California at Santa Barbara. He is a member of several scientific societies, including the Geological Society of America, the American Geophysical Union, and the National Ground Water Association.

LYNN W. GELHAR is a professor of civil and environmental engineering at the Massachusetts Institute of Technology. His research activities encompass stochastic theories of transport processes for unsaturated flow, fractured media, chemically heterogeneous media, variable viscosity fluids, biodegradation, multiphase flow, controlled field experiments on macrodispersion in aquifers and unsaturated flow, supercomputer simulation of flow and transport in heterogeneous porous media, and stimulation of in situ biodegradation using gas injection. Dr. Gelhar has more than two decades of experience on aspects of subsurface hydrology relating specifically to problems of radioactive waste disposal in the United States and abroad. He has served on several multidisciplinary review teams, including groups reviewing environmental aspects of the Hanford site, the Waste Isolation Pilot Plant (WIPP) site, and the Nevada Test Site. He is currently serving on the NRC Panel on Conceptual Models of Flow and Transport in the Fractured Vadose Zone. Dr. Gelhar holds B.S., M.S., and Ph.D. degrees in civil engineering from the University of Wisconsin.

LISA C. GREEN is a technical manager with Lucent Technologies in Norcross, Georgia. Her expertise is in chemical engineering and analytical characterization applied to large-scale manufacturing, and she has more than 20 years of process engineering and project management experience. At Lucent, she has had extensive experience in engineering and management of chemical processing and waste management systems, including experience in working with regulatory agencies, risk management teams, and insurance companies. She holds an M.S. degree in analytical chemistry and a B.Che. degree from the Georgia Institute of Technology. She has received three STAR (Significant Technical Achievement Recognition) awards and three Lucent Environmental Hero Awards for her work.

Biographical Sketches

ROBERT O. HALL, JR., is an assistant professor in the Department of Zoology and Physiology at the University of Wyoming. His current research interests include interactions of aquatic community structure and ecosystem function, energy, and nutrient flow in food webs; bacterivory by aquatic invertebrates; stable isotopes as food web tracers; and nitrogen cycling. His current research projects include estimating controls of nutrient uptake and retention in streams in Grand Teton National Park. Dr. Hall holds a B.S. degree from Cornell University and a Ph.D. degree from the University of Georgia.

EDWIN E. HERRICKS is professor of environmental biology in the Department of Civil and Environmental Engineering at the University of Illinois at Urbana-Champaign. His areas of expertise include aquatic ecology and stream ecosystem and watershed management, and he has broad experience in the identification, assessment, and restoration of the adverse effects of man's activities on streams, rivers, lakes, and their watersheds. His current research has focused on the development of methods to restore and manage wetland design and management, the development of test systems to assess episodic exposure to contaminants common in urban runoff, and the assessment of the effects of global climate change on natural resources. Dr. Herricks is currently serving on the NRC's Surface Transportation Environmental Research Advisory Board. He has written numerous articles and papers on the broad theme of improving engineering design and environmental decision making. He is a member of the Urban Water Resources Research Council of the American Society of Civil Engineers and chairman of a task group on receiving system effects from urban runoff. He holds a B.A. in biology and english from the University of Kansas, an M.S. in engineering from the Johns Hopkins University, and a Ph.D. in biology from the Virginia Polytechnic Institute and State University.

BRUCE D. HONEYMAN is a professor of environmental science and engineering at the Colorado School of Mines in Golden. His research interests include the physical and chemical processes controlling the fate of chemical species, especially radionuclides, in natural and engineered systems; nuclear environmental chemistry; actinide surface chemistry; colloid-facilitated contaminant transport; metal-organic complex formation; and treatment of radionuclide-contaminated environmental media. Dr. Honeyman holds a B.S. degree in applied earth sciences and M.S. and Ph.D. degrees in environmental engineering and sciences from Stanford University. He is a member of several professional societies, including the American Chemical Society, the American Geophysical Union, and the American Society of Limnology and Oceanography.

SALOMON LEVY (NAE) is a retired chairman and chief executive officer of S. Levy Inc., a consulting firm to the power industry that he formed in 1977. He is now a principal in Levy & Associates and continues to provide personal consulting services to the power industry. He has served as a consultant or on oversight committees for several utility companies and as an adjunct professor at the University of California, Berkeley and Los Angeles. From 1953 to 1977, the General Electric Company employed him in a variety of technical and managerial positions, the last of which was as general manager of the Boiling Water Reactor Operations. He is a member of the National Academy of Engineering and has served on several NRC committees. Dr. Levy holds B.S., M.S., and Ph.D. degrees in mechanical engineering from the University of California at Berkeley.

JAMES K. MITCHELL (NAS, NAE) is University Distinguished Professor Emeritus at Virginia Polytechnic Institute and State University in Blacksburg and a consulting geotechnical engineer. Dr. Mitchell's expertise is in civil engineering and geotechnical engineering, with emphasis on problems and projects involving construction on, in, and with the earth; mitigation of ground failure risk; waste containment and site remediation soil improvement; soil behavior; geotechnical earthquake engineering; environmental geotechnics; and compositional and physicochemical properties of soils. He has served on several National Research Council committees, most recently the BRWM-Water Science and Technology Board (WSTB) Committee on DOE Research in Subsurface Science. Dr. Mitchell holds a B.S. degree in civil engineering from Rensselaer Polytechnic Institute, and M.S. and Ph.D. degrees from the Massachusetts Institute of Technology.

LEON T. SILVER (NAS) is the W.M. Keck Foundation Professor for Resource Geology (emeritus) at the California Institute of Technology. Dr. Silver's research focuses on several areas of geology and geochemistry including the petrology and history of the continental lithosphere. He has served on numerous National Research Council committees, most recently the BRWM-WSTB Committee on DOE Research in Subsurface Science. Dr. Silver was elected to the National Academy of Sciences in 1974. He holds a B.S. degree in civil engineering from the University of Colorado, an M.S. degree in geology from the University of New Mexico, and a Ph.D. degree from the California Institute of Technology.

LESLIE SMITH is the Cominco Chair in Minerals and the Environment at the University of British Columbia in Vancouver. His expertise is in the areas of subsurface hydrology and contaminant transport processes. His

Biographical Sketches 167

current research interests include transport processes in fractured rock masses, hydrologic processes in unsaturated waste rock piles, hydrogeological decision analysis and risk assessment, inverse modeling, and radionuclide transport in watersheds near the Chernobyl Nuclear Power Plant in the Ukraine. Dr. Smith has served on several National Research Council committees, including the BRWM's Committee to Review Specific Scientific and Technical Safety Issues Related to the Ward Valley, California, Low Level Radioactive Waste Site. Dr. Smith holds a B.S. degree in geology from the University of Alberta and a Ph.D. in geological sciences from the University of British Columbia.

DAVID A. STONESTROM is a hydrologist with the U.S. Geological Survey (USGS) in Menlo Park, California. He is chief of the research project titled "Application of Unsaturated Flow Theory to the Phenomena of Infiltration and Drainage" for the National Research Program of the Water Resources Division. He is also a coordinator of the Amargosa Desert Research Site for the USGS Toxic Substances Hydrology Program. His current research investigates processes governing the occurrence and movement of gases and liquids in unsaturated zones by applying principles of soil physics, pedology, and geochemistry. He holds a B.S. degree in geology from Dickinson College and M.S. and Ph.D. degrees in hydrology from Stanford University. Dr. Stonestrom holds membership in several professional societies, including the American Geophysical Union, the Soil Science Society of America, and the Coalition for Earth Science Education.

B
Information-Gathering Meetings

Presentations Given During First Committee Meeting
(April 11-12, 2000, Richland, Washington)

Background (Gerald Boyd, Deputy Assistant Secretary, U.S. Department of Energy [DOE], Office of Science and Technology [via telephone])

Hanford Site history (Roy Gephart, Pacific Northwest National Laboratory [PNNL])

Hanford Site vision and future (Mike Thompson, Acting Program Manager, Groundwater/Vadose Zone Office, DOE Richland)

Overview of the Groundwater/Vadose Zone Integration Project (Michael Graham, Bechtel Hanford)

Overview of the Hanford Site science and technology (S&T) program (Mark Freshley and John Zachara, PNNL)

Overview of the activities of the Integration Program Expert Panel (IPEP) (Ed Berkey, Concurrent Technologies Corporation, IPEP Chair)

Comments from stakeholders, regulators, and Tribal Nations

Presentations Given During Second Committee Meeting
(June 28-30, 2000, Richland, Washington)

What is the end-state vision for the Hanford Site, and what decisions need to be made to achieve this vision? (Harry Boston, DOE Richland; Mike Hughes, Bechtel-Hanford)

Discussion of R&D needs and S&T plan (John Zachara and Mark Freshley, PNNL)

Comments from stakeholders, regulators, and Tribal Nations

Presentations Given During Third Committee Meeting
(September 6-8, 2000, Richland, Washington)

River, groundwater-river interface, and risk (Roger Dirkes, Amoret Bunn, Integration Project [IP])

Characterization of systems and inventory (includes inventory, monitoring, characterization, and data management) (Bruce Ford, Charley Kincaid, IP)

System Assessment Capability: Definition and development and description of current activities and future plans (Bob Bryce, IP)

Comments from stakeholders, regulators, and Tribal Nations

Field Trip to Hanford Site
(Wednesday, September 6, 2000)

Gable Mountain:
Geology-Hydrogeology
Nuclear fuel cycle, operations, and waste disposal history
Cleanup plans

100-H Area:
Decontamination and decommissioning (D&D); environmental restoration activities; Columbia River and salmon spawning grounds.

100 D Area:
Chromium plume and in situ redox passive barrier

100 N Area:
Strontium-90 plume and treatment operations
Field Lysimeter Test Facility
Z trenches (carbon tetrachloride site)
SX Tank Farm
Environmental Restoration Disposal Facility (ERDF) (Hanford sand facies and clastic dikes)
B, BX, and BY Tank Farms (single shell)
BY Cribs (200-BP-1)
Hanford Engineered Barrier (Hanford Cap)
Vadose zone transport field study site (Sisson and Lu site) and ILAW

Presentation Given During Fourth Committee Meeting
(November 1-3, 2000, Irvine, California)

Update on the integration project (Michael Graham, Bechtel Hanford; Mark Freshley, PNNL)

Fifth Committee Meeting
(January 18-19, 2001, Irvine, California)

No presentations were given at this meeting.

Presentations Given During Sixth Committee Meeting
(March 28-30, 2001, Washington, D.C.)

Integration Project update (Michael Graham, Bechtel Hanford)

S-SX field investigations (Tony Knepp, Office of River Protection)

Science and technology contributions to S-SX field investigations (John Zachara, PNNL)

C
Scaling Issues Applicable to Environmental Systems

The essential problem with using models to predict the behavior of environmental systems is that the scale of interest for predictions is rarely, if ever, the scale for which information is available to construct and validate the model. As a consequence, projections in time and space must be made often without needed validation at the target scale. The goal of *scaling* is to capture essential system characteristics at a scale of direct observation and to extrapolate to a different scale. Although all environmental systems present scaling problems, the natural heterogeneity of the subsurface environment requires model predictions of contaminant transport over spatial scales that may range from the "grain" scale of several millimeters to field scales of kilometers; in addition, temporal scales may require accommodating the simulation of events that require hundreds or thousands of years to complete (e.g., the dissolution of minerals). Predictions of contaminant behavior at scales of interest to environmental managers is currently problematic because of a general lack of understanding of both theoretical and applied approaches to scaling environmental phenomena.

Figure C.1A-C illustrates some of the scaling issues at the Hanford Site. One scale length of importance at Hanford is the site itself. An example of a problem at this scale is the need to quantify the potential effect that 200 Area contaminants will have on Columbia River water quality. This scale length is shown schematically as bar *a* in Figure C.1A and C.1B. The site scale is at the upper limit of length scales in this analysis. Other environmental scales of interest to groundwater modeling include the vertical and horizontal extent of a lithologic unit, *c* and *e*, respectively; the vadose zone thickness, *d* (which is itself a complex hydrologic environment; see Chapter 6); and the scale of individual minerals and colloids, *b*. Given these differences, a further complication in scaling is the development of an accurate understanding of the scale length of a portion of the target system that can be considered to be homogeneous with respect to geochemical and hydrologic properties (*f*). It is possible to determine the scale length of such representative units, but the scale is dependent on location in the environmental system and the time of system evolution.

Scales of observation for experiments, which are used to develop models, rarely conform to the environmental scale of interest to the environmental manager. At one end of the spectrum are observations of system behavior and characteristics at the molecular to grain scale (*g, h*),

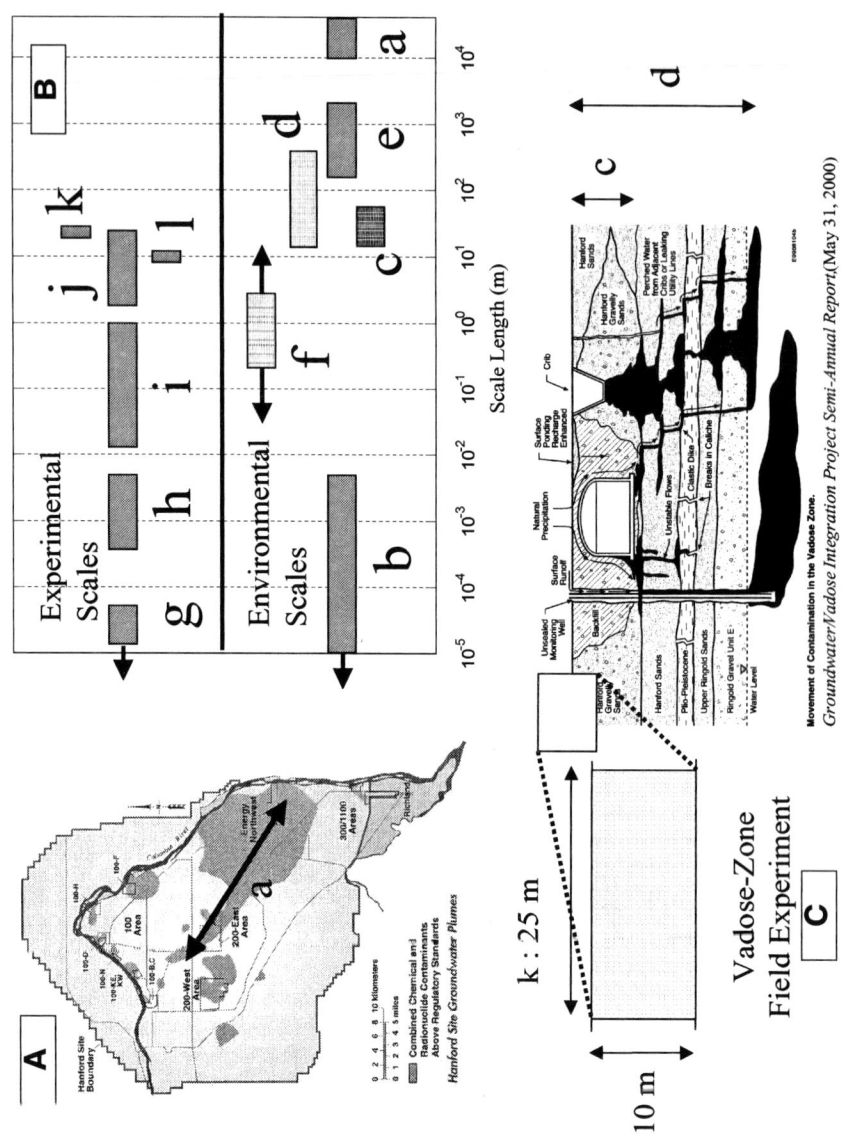

Figure C.1. Comparison of environmental and experimental scales at the Hanford Site. (A) Scales (>20 km) considered by the Systems Assessment Capability (SAC). (B, C) Scale-length comparisons. Bars show approximate scales of (*a*) the Hanford Site modeled by SAC; (*b*) mineral grains, microbiota, and colloids; (*c*) thickness of hydrogeologic units; (*d*) vadose-zone thickness; (*e*) lateral extent of hydrogeologic units; (*f*) model discretization (i.e., assumed hydrologic and chemical homogeneity); (*g*) colloidal studies; (*h*) mineral grain studies; (*i*) bench-scale (core and beaker) laboratory experiments; (*j*) larger-scale laboratory experiments; (*k*, *l*) horizontal and vertical dimensions of the vadose-zone field experiment.

which are essential for understanding fundamental processes and providing confidence that models capture essential processes. Bench-scale experimentation (*i*) may capture the geochemical or hydrologic properties of meter-scale systems, but such information does not readily scale up to larger systems (*j*) of 10 meters in size, for example. Part of the problem in scaling geochemical processes is a lack of understanding of the nature of geochemical heterogeneity and the ways in which the *distribution* of heterogeneity affects processes at different scales. The vadose zone field experiment at Hanford (Ward and Gee, 2000) is a proposal to evaluate in situ properties. Despite its proposed scale (Figure C.1C: *l* and *k*), it is unclear whether the test bed will capture the complexity of the vadose zone sufficiently well that extrapolation to other scales of interest (*c-e*) will be possible.

UPSCALING TRANSPORT BEHAVIOR

Because of spatial variability in the subsurface and the time required for of physical and chemical processes to occur in groundwater, it is not possible to use measured transport properties from a few laboratory experiments to model field-scale behaviors accurately. A key parameter in any model of groundwater movement is permeability. Permeability is important because it determines the potential speed of contaminant migration associated with subsurface water movement. Permeability is observed to vary by several orders of magnitude over distances as small as meters in a given geologic unit. Variations of permeability and other transport properties occur over scales of fractions of a meter, making it impractical to completely map out these characteristics at the field scale of interest. Without this detailed mapping, exact predictive modeling is problematic.

Stochastic characterization of the spatial variability of transport properties has been found to be an effective way to treat subsurface heterogeneity and to represent transport properties at the field scale. For example, for a nonreactive solute in a saturated aquifer, variations in permeability cause variations in velocity that produce spreading of contaminants relative to the bulk flow. Stochastic analyses describing the variations in velocity are used to derive the mean transport equation that represents the large-scale transport process and the transport parameters, such as macrodispersivity, appearing in the large-scale transport equation.

Scaling Issues

The stochastic upscaling approach has been developed extensively over the past 20 years (see Dagan, 1989; Gelhar, 1993; 1997) and has been tested in field experiments demonstrating that field-scale transport properties, such as macrodispersivity, could be predicted independently by carefully designed measurements of the logarithm of permeability covariance. This stochastic upscaling approach has the advantage that it provides a systematic framework through which feasible small-scale laboratory or field measurements of medium properties can be used to predict large-scale transport properties, thereby showing explicitly how additional data will improve the estimated large-scale transport properties. This approach also has the advantage that it can be used to quantify the uncertainty in large-scale predictions by evaluating the concentration variance as a measure of the variation around the mean solution. A disadvantage of the stochastic upscaling approach is the extensive, statistically focused data requirements; standard site characterization efforts typically do not provide the type of data required to implement this approach.

Vadose zone transport processes are influenced strongly by natural heterogeneity in the subsurface environment because of the nonlinear nature of unsaturated flow (see Chapter 6). Permeability in an unsaturated system depends on both the medium and the fluid. Stochastic upscaling treatments have been applied to unsaturated systems and show that layered heterogeneity can strongly enhance horizontal moisture movement under low-moisture conditions (Yeh et al., 1985). The data requirements for stochastic upscaling in the unsaturated zone are more severe because of the difficulties of measuring unsaturated properties accurately and efficiently in the laboratory. Transport properties such as macrodispersivity are, in principle, predictable via stochastic upscaling (Gelhar, 1993, p. 261; Russo, 1997), but this approach has not been tested under field conditions.

Heterogeneity of chemical properties and the relationship to flow properties can have an important influence on large-scale transport properties for reactive contaminants in the vadose zone. Stochastic analyses and numerical simulations show that variations in retardation factors and their relationship to permeability can significantly increase the macrodispersivity of the sorbed contaminant relative to that of the nonsorbing species (Gelhar, 1993, p. 256; Talbott and Gelhar, 1994). To determine the relationship between chemical and flow properties, sampling programs for describing reactive transport characteristics must be designed carefully to ensure that both chemical and flow properties are determined for individual samples.

GEOCHEMICAL HETEROGENEITY AND SYSTEM SCALING

A long-standing problem in subsurface transport modeling has been the accurate description of chemical processes regulating contaminant retardation. The development of sorption models explicitly considering physicochemical processes that produce accumulation of ions at interfaces has relied on the analysis of well-characterized monomineralic systems with the often implicit assumption that overall system behavior can be described through a "summation" of component behavior (e.g., Honeyman, 1984). However, interactive effects confound this approach. Alternatively, it is possible to incorporate surface chemical models using the Generalized Composite Method (Davis et al., 1998) in which the representative geomedia is treated as an undifferentiated whole.

Both the explicit consideration of interfacial processes in regulating contaminant retardation and the role of permeability distribution in affecting macrodispersivity rely on the development of a means of representing the distribution of heterogeneity. Considerable work has been done on the distribution of permeability in geomedia of different scale lengths and its contribution to solute transport. Lagging far behind, however, is the development of an understanding of the distribution of geochemical parameters (i.e., the heterogeneity field) that regulates the retardation of surface-active contaminants. In either case, the ability to scale up from well-defined systems to scales of environmental interest requires sampling and system characterization campaigns designed specifically to capture the uncertainty in such a manner as to adequately bridge the scales.

SCALE ISSUES AND WATERSHEDS

As an environmental system, river channels also present important scaling issues, and like the subsurface environment, river channels have boundaries defined in both space and time. In analyzing these dimensional issues with reference to water quality, the primary issue is the importance of length scales, which range from microscopic to watershed scales of hundreds of kilometers. In addition to water quality, channel characteristics, which constrain habitat for aquatic organisms, also present scale issues in both time and space domains.

Geomorphologists have developed an effective method of working with temporal scales in channel networks. Schumm and Lichty (1965) proposed a conceptual framework for geomorphological time that classifies temporal dynamics and changing relationships among system variables when temporal scales are traversed. Although the three temporal modes—

cyclic, graded and steady—do not have an absolute value, they differ in duration. Cyclic time is long, related to an erosion cycle of uplift and erosion to some level, a time frame associated with landscape change. Graded time is a short period of cyclic time that is associated with graded river profiles that represent periods of channel stability over hundreds to thousands of years. Steady time is short and applies to processes that occur along a river reach in seconds to years. With these definitions of temporal scales, Schumm and Lichty provided a system that relates channel and flow variables in each time mode by considering how independence-dependence relationships among variables change with the perspective of time scale. Independence-dependence is important in any analysis because the independent variable (the cause) will be the controlling factor in an analysis by producing a response in the dependent variable.

Following the Schumm and Lichty approach, it is possible to consider spatial and temporal scales in analyzing water quality issues in watersheds that include both chemical water quality and physical conditions that define habitat. At large spatial and temporal scales, the emphasis of analysis will be on source development and contaminant loading. At smaller spatial scales, the emphasis will be on concentration and duration of exposure. Further, spatial and temporal scales define external factors that relate dependent to independent variables and, most importantly, cause and effect. As scales of analysis are reduced, greater numbers of environmental and water quality variables can be considered independent, leading to a better definition of cause and effect, and directing management efforts to specific actions.

Frissell et al. (1986) proposed a habitat-centered view of stream systems, a modified version of which is shown in Table C.1. Their view is based on a hierarchical organization of habitat types. In this hierarchical organization, subsystems (stream segments, reaches, pools or riffles, and microhabitats) develop and persist within a specified spatiotemporal scale. In this "systems" view, high-frequency, low-magnitude geomorphic events of the steady time scale predominate in subsystems, while the system as a whole is subject to low-frequency, high-magnitude events of graded or cyclic time scales. A critical issue in the hierarchy, particularly when considering water quality issues, is that the setting within which components, process, and dynamics are defined is provided by the next-higher level in the hierarchy. These "nested" relationships in the hierarchy provide an example of the integration of Schumm and Lichty's time-scale perspective, illustrating the change in dependence relationships at different levels of the hierarchy. Recognition of this change in controlling variables with time-scale perspective is particularly important in the management of riverine ecosystems.

TABLE C.1. Expected Spatial and Temporal Scales, Events, Processes, and Water Quality Issues

Level in Watershed Hierarchy or Scale	Spatial Scale (km)	Time Span	Expected Duration (years)	Landscape Change Characteristics	Process Characteristics	Water Quality Issues
Stream system or basin	$10^3 - 10^4$	Cyclic-geologic	10^6-10^5	Tectonic uplift, subsidence, sea-level changes, glaciation	Planation denudation drainage network development	Source materials established in watershed; weathering products dependent on parent materials and exposure conditions
Segment system or network	10^2-10^3	Cyclic-geologic	10^4-10^3	Minor glaciation, earthquakes, very large landslides, alluvial or colluvial valley infilling; river meander pattern development	Migration of tributary junctions and bedrock nickpoints, channel downcutting, extension of first-order channels	Periodic exposure of unweathered materials may lead to water quality differences in basin; expectation of mid- to late-stage weathering products; more complete channel network increases potential for landscape change to affect water quality
Reach system or channel reach	10^1-10^2	Graded-modern	10^1-10^2	Debris torrents; landslides; channel shifts; river meander cutoff; channelization, diversion or damming by man	Aggradation-degradation associated with sediment storage structures bank erosion riparian vegetation succession	Loading emphasis; focus on input, output, and residence time of weathering products and contaminants; adjustment of channel form to sediment loading and corresponding storage of contaminated materials; exposure times changing from chronic to acute
Pool-riffle system, channel cross section	10^0	Steady-present	10^1-10^0	Sediment accumulation or washout; small bank failures; floodflow scour or deposition; thalweg shift; alternate bar change	Small-scale lateral or elevational changes in bedforms	Shift to concentration emphasis; presence of "hot spots" of contamination; water column–dissolved contaminants of greater importance; emphasis on acute exposure levels
Microhabitat system	~0^{-1}	Steady-present	$10^0 - 10^{-1}$	Annual flow, sediment and organic matter transport; substrate scour; seasonal macrophyte or periphyton growth and scour	Seasonal depth, velocity changes, sediment accumulation, biological processes	Concentration and duration variability, high; mixing zone issues, important; emphasis on less-than-acute exposures that emphasize return frequency and timing of water quality alteration

SOURCE: Modified from Frissell et al., 1986.

REFERENCES

Dagan, G. 1989. Flow and Transport in Porous Formations. Berlin: Springer-Verlag.

Davis, J.A., J.A. Coston, D.B. Kent, and C.C. Fuller. 1998. Application of the surface complexation concept to complex mineral assemblages. Environ. Sci. Technol. 32: 2820-2828.

Frissell, C.A., W.J. Liss, C.E. Warren, and M.D. Hurley. 1986. A hierarchical framework for stream habitat classification: viewing streams in a watershed context. Environmental Management, 10: 199-214.

Gelhar, L.W. 1993. Stochastic Subsurface Hydrology. Englewood Cliffs, N.J.: Prentice Hall.

Gelhar, L.W. 1997. Perspectives on field scale application of stochastic subsurface hydrology: in Subsurface Flow and Transport: The Stochastic Approach, G. Dagan and S.P. Neuman, eds., Cambridge University Press/UNESCO.

Honeyman, B.D. 1984. Cation and Anion Sorption in Binary Mixtures of Adsorbents: An Investigation of the Concept of Adsorptive Additivity. Ph.D. thesis, Stanford University.

Russo, D. 1997. Stochastic analysis of solute transport in partially saturated heterogeneous soils: in Subsurface Flow and Transport: The Stochastic Approach, pp.196-206, G. Dagan and S. P. Neuman, eds., Cambridge University Press/UNESCO.

Shcumm, S.A., and R.W. Lichty. 1965. Time, space and causality in geomorphology. American Journal of Science, 263: 110-119.

Talbott, M.E., and L.W Gelhar. 1994. Performance Assessment of a Hypothetical Low-Level Waste Facility: Groundwater Flow and Transport Simulation. Washington, D.C.: U.S. Nuclear Regulatory Commission Report NUREG/CR-6114, Vol. 3.

Ward, A.L.. and G.W. Gee. 2000. Vadose Zone Transport Field Study: Detailed Test Plan for Simulated Leak Tests. Richland, Wash.: Pacific Northwest National Laboratory. PNNL-13263.

Yeh, T.-C., L.W. Gelhar, and A.L. Gutjahr. 1985. Stochastic analysis of unsaturated flow in heterogeneous soils. 3. Observations and applications. Water Resources Research 21(4): 465-471.

D
Acronyms

AEA	Atomic Energy Act (1954)
CERCLA	Comprehensive Environmental Response, Compensation, and Liability Act
CFR	Code of Federal Regulations
DNAPL	dense non aqueous phase liquid
DOE	U.S. Department of Energy
DQO	Data Quality Objectives
DST	double-shell tank
DWPF	Defense Waste Processing Facility (Savannah River vitrification plant)
EM	Office of Environmental Management
EMSP	Environmental Management Science Program
EPA	U.S. Environmental Protection Agency
ERDF	Environmental Restoration Disposal Facility
FY	fiscal year
GAO	U.S. General Accounting Office
HDW	Hanford Defined Wastes
HEIS	Hanford Environmental Information System
HLW	high-level waste
INEEL	Idaho National Engineering and Environmental Laboratory
IPEP	Integration Project Expert Panel
IRIS	Integrated Risk Information System (EPA)
LLW	low-level waste
NRC	National Research Council
PNNL	Pacific Northwest National Laboratory
PUREX	Plutonium and Uranium Extraction
RCRA	Resource Conservation and Recovery Act
ROD	record of decision
SAC	System Assessment Capability
SST	single-shell tank
S&T	Science and Technology
STCG	Site Technology Coordination Group
SWITS	Solid Waste Inventory Tracking System
TCD	Tank Characterization Database
TRAC	Track Radioactive Component
TWRS	Tank Waste Remediation System
USGS	U.S. Geological Survey
WIDS	Waste Inventory Data System
WIPP	Waste Isolation Pilot Plant
WSTB	Water Science and Technology Board